DO NO HARM

THE DIARY OF A 'VACCINATED' ANTI-VAXXER

DR DAVID CARTLAND

BMEDSCI MBCHB

Fisher King Publishing

DO NO HARM
THE DIARY OF A 'VACCINATED' ANTI-VAXXER

Print ISBN 978-1-916776-79-1
Ebook ISBN 978-1-916776-80-7

Published worldwide by Fisher King Publishing
fisherkingpublishing.co.uk

Printed and bound by CPI Group (UK) Ltd, Croydon, CR0 4YY.

FSC

I dedicate this book to all those brave medics, scientists, nurses and souls that had the courage to stand up, speak out and risk it all, in their own way, for the sake of humanity in the face of mass medical negligence.

I also dedicate this book to every victim of the mRNA covid gene technology. I look forward to a time that justice will be served, the gaslighting ends, and that help, and support is finally given to those who were injured and to the bereaved families of those who have been killed.

Finally, to my children, and my eventual grandchildren, this book is an accurate record of the part your dad played towards awakening humanity rather than the distortion of the facts created by the tabloids and the misinformation presented by the regulators. This is for you and for generations to come.

Acknowledgments

There are far too many people to thank for the space available here but to all those who inspired me to write this book and who encouraged me to tell my side of the story, I will never forget your unstinting support.

Huge thanks to Ian McDermott, for being there to listen at the darkest of times, I will be forever grateful.

Thanks to Ros Jones for her 'above and beyond' help on so many fronts and her continual support and wise counsel.

Thank you too goes to Mike Fairclough, a brave warrior who not only helped to edit these pages but whose courage was an enormous inspiration in my decision to speak out. The lone wolf among his profession inspired my awakening as he has inspired so many around the world.

To Rick Armstrong, Rich Hayden, and all at Fisher King Publishing for believing in me, while steadfastly being there to help me turn an idea into a reality.

A special mention to Jane Angel, a great friend to have, especially in times of adversity, for her constant help, words of wisdom, practical support and ongoing encouragement.

And finally, to my wife, you are my rock and my best friend in this world. You have stuck with me through unprecedented pressure and in spite of a level of multifaceted stress that would have made lesser mortals buckle. I love you.

Renowned cancer geneticist says his lab uncovered that mRNA shots 'Permanently Damage Human Genome.'
Phillip Buckhaults, PhD, Cancer Geneticist

We see a constellation of symptoms, that is not long covid, IT IS LONG VAX. One of these injuries, Myalgic encephalomyelitis/chronic fatigue syndrome (ME/CFS), is caused by the spike protein. Young healthy people's lives are destroyed.
Dr Pierre Kory, Chief Medical Officer at Front-Line COVID-19 Critical Care Alliance

mRNA technology has filled my clinic with patients asking, 'How long will I have this problem?'
Professor Robert Clancy, Clinical Immunologist

mRNA vaccines should never be put in a human body.
Florida Surgeon General Dr. Joseph Ladapo

The mRNA platform is fatally flawed, dangerous and inferior to natural immunity.
Prof. Brett Weinstein, Evolutionary Biologist

From the data of 21 million of Japan's vaccinated we discovered a disturbing spike in deaths between 90 and 120 days after mRNA vaccination, resulting in the deaths of more than 600,000 people.
Professor Murakami, University of Tokyo

Covid vaccines are neither safe nor effective.
Joshua Rys, Johnson & Johnson's chief scientific officer of regulatory affairs

mRNA vaccines may trigger genetic chaos.
Dr. John Catanzaro, Integrative Oncologist

It is very likely that mRNA vaccines accelerate coronary artery disease.
Assem Malhotra, Cardiologist

Contents

Introduction

The pandemic years (which I will refer to as 'scamdemic' from this point on), December 2019 to the present day, has become a living nightmare for anyone unrelated to the pharmaceutical industry and its kickbacks. It has left in its wake financial ruin, a mental health disaster instigated by an orchestrated campaign of psychologic operations, healthcare trust decline, and science leaving the building with its mate's evidence-based medicine and biomedical ethics, to name but a few.

Thanks to the sheer 'fear porn,' frivolous and negligent 'problem-reaction-solution' approach made in world-wide lockstep we are now faced with a very real pandemic of sorts. Broken immune systems, cancer, chronic inflammation, and autoimmunity as well as cumulative micro/macrovascular clotting and aberrant protein deposition throughout the human body. The consequences of taking the covid-19 jabs and its multifaceted pathophysiology wreaking havoc in this novel form of gene-based technology never before used outside of clinical trials along with nanotechnology. The public, unbeknown to them, had agreed by rolling up their sleeve to become a participant in the biggest clinical trial in history while not realising their role as proverbial lab rats until the (unaspirated) syringe was plunged! Or was it indeed a clinical 'trial' or a premeditated plan?

The ‘solution’ has left in its wake biological carnage in the form of multitudes of individuals dead and disabled who trusted the very people they placed their lives in the hands of: the government, healthcare staff, and scientists to find the answer to the ‘deadly’ scamdemic of coronavirus given the special Hollywood billing of ‘Covid-19’ to delineate it from the many thousands of circulating self-limiting coryzal illnesses that this naturally circulating seasonal virus the form of which it usually takes.

But enter stage left the media propaganda machine guided by the owners of the MSM and their ‘sponsors,’ ‘Big Pharma,’ to catastrophise events with footage of packed ITU’s, hazmat suit clad, goggle-wearing, gaunt, emotional staff and material beamed ‘live from Wuhan/Bergamo’ with the dramatic terrible effect of sudden collapses in the street invariably caught on camera for this respiratory virus. The daily briefings began soon after and the narrative was firmly in control of the powers that be. The media and pharmaceutical industry pitched the solution from this deadly scourge, promoted as the holy saviour of this ‘zombie apocalypse,’ the covid-19 ‘vaccine.’

Being a dissident voice and in particular a medical voice in this one-sided paradigm of deadly disease/safe and effective has turned out to be career ending for the brave few who stood up for patient safety and biomedical ethics and who shone a light on evidence-based practice, keeping up with world's mortality and morbidity data and emergent studies. And those who held firm to their Hippocratic oath of 'Primum non nocere,' first do no harm, were soon subject to disciplinary action and professional cancel culture. These terrible 'anti-vaxxers' were not to be tolerated by the medical world, like lambs to the slaughter, picked off as individuals by regulators, kept silent in the media and social media through nefarious censorship and organisational targeting.

In this book I seek to document my version of events. On the 'frontline,' across many practices and arenas, not as an undercover agent of sorts but a doctor who took his oath and professional obligations seriously and remembered from day one why he chose to become a medic. I had absolutely nothing to gain from speaking out about my concerns and everything to lose (I can firmly attest to the latter). A version of events which have never been allowed to be made public until recent easing of social media control and removal of censorship and the drip feed of mainstream media slowly allowing the chaos that has ensued led by our worlds governments and medical profession. This has left a trail of death, disability, destruction, and a broken trust that will never be fixed, unless a groundswell of medics and nurses decide to leave behind their material belongings, forget the lucre offered to stay silent, and remember why they entered the profession to begin with. Most importantly, the DUTY of candour, set up by healthcare to maximise confidence and trust in this once noble profession.

I will keep this commentary/diary of events close to my personal experience and refrain from tangentially leaving my own clinical and scientific experiences. The pen is mightier than the sword!

Chapter 1 - Why did I become a doctor?

As an introduction, let me take you back to the morning of my medical school interview. I was riddled with anxiety as had spent the last twenty-three years in one way or another preparing for the questions that I would be asked to enter into this hallowed and highly regarded profession. Medicine, with such rich history, was so highly and universally respected. I had done a previous degree at the same institution achieving first class honours in Biomedical sciences and having not managed to get into the new 'graduate entry' course at first attempt (the medical school had opted for an experimental approach admitting graduates from non-medical first degrees). After a brief time back at my A-Level school to try and achieve an award for Chemistry, suggested by the admissions tutor, opted to leave and start an MSc in medical research that followed on nicely from my dissertation in Angiogenesis where I was successful in authoring published research into medical journals. One thing I took from those years in research was the instilling of a critical mind, always question, look for conflicts, review all claims and apply logic consistently. This is in stark contrast to the more spoon-fed approach of learning the medical curriculum. I was taught to learn about root causes of disease and pathophysiology, but this was sparser and a focus more on the medical curriculum's rote learning ad-verbatim methods.

Anyway, back to the interview. I had mulled over how I would answer the obvious question and had several incarnations of an answer, some more creative than others. I was all rehearsed and ready for this question as I tentatively walked into the room facing a panel and my arch nemesis, the admissions tutor, who I had eagerly chased around the medical school for the best part of four years to get this opportunity. Braced for the question 'Why do you want to be a doctor?' After introductions, I was completely hoodwinked by the initial enquiry. I was asked what my parents did for a living! My stomach sank as I was thrown. I am a working class Brummie lad who happened to have an enquiring mind, photographic memory, and a nuance for all things science, but my upbringing was not one associated with the private school backgrounds and affluence that the profession traditionally attracted mainly due to the cost of training. Without sounding like our latest WEF preferring Prime Minister, my dad was a toolmaker, and my mom worked part time in retail. I had two brothers, and I was the middle child; we had what I would call a 'proper 90's' upbringing filled with outside living, lots of friends, playing sport and being out until dusk. We just about got by and looking back now, I fully appreciate the sacrifices my parents made to put food on the table and keep us clothed and looked after. A proper working-class family unit.

Onto the next question and, completely stumped by its predecessor, I was asked whether I thought medicine was an art or a science. I think I gave an answer somewhere in the middle ground and would agree with that response to this day. Then it came, the question for which I had polished an answer in my mind. Why do you want to become a doctor, David? After a short pause, my mind went blank, and my previous premeditated answer evaporated. I was left in that moment to speak from my heart, palms sweating and voice shaky. I cannot remember the exact wording, but it focussed entirely on my deep-seated wish and hope to live my life helping people through their physical, mental, and sociological ill health and distresses. It wasn't as much a hope than obsession given my whole life had been geared to doing the degree to enable me in this passion. I added more depth to the response, but it was all about caring for my patient in their hour of need. That is the crux of why I gave my life to this profession. I stand by my reply to this day and often go back to this response when people enquire about why I spoke out during the pandemic. My patients. Primum non nocere. My willingness to help people heal and hear them in their time of need and vulnerability.

I had a great time at medical school for the four years I trained. This was split into a year of preclinical work focussed on problem-based learning, a new concept where responsibility to learn the curriculum was on the students to research and teach each other based around a clinical scenario. The dropout rate was high from this second intake due to the intensity of training and regular need for exams especially the dreaded 'OSCE,' objective structured clinical examination. I managed to use every ounce of resilience I had to get through that year and remember the pressure of 2004-2005 extremely well. The 'survivors' went on to join the third year mainstream intake to do the clinical rotations that took me around Birmingham hospitals and general practice to learn my trade. I loved my initial patient interactions and remember well those early positive feedback moments as well as the fine tuning of the art of taking a history, examining, organising investigations to diagnose and subsequently heal patients. I also remember those sad moments and while not directly involved in the patients care as a trainee, I still remember the faces of patient and relatives as I witnessed patients succumb to their injury or illness.

I remember navigating the pathway to finals exams as I prepared meticulously and was confident of becoming a fully-fledged medical doctor at what ended up being an eight year extremely expensive stint at medical school. I recall results day like it was yesterday as I opened the envelope to find that as of that moment I was to be known as Dr David James Cartland. I even managed to get honours in a couple of modules. If I didn't have a sternum, my heart would have burst out of my chest. The pride, the relief, the release of emotion is something I will take to my grave. But finally, I was on the road to directly managing and helping my patients on their journey back to health. This moment was only superseded by graduation day. As a surviving bunch of graduate entry medics, we had always remained close despite being mixed in with the mainstream and had always gravitated toward each other in and out of medical school. Group resilience. I sat near many of them when the time had come to collect that certificate which had cost me blood, sweat and tears, in addition to a substantial student loan. I proudly marched up to take my certificate shaking the hand of the Chancellor firmly and confident as my mother and father looked on. Dr David Cartland was born, the first ever in his family, the quiet nervous chap from his first degree who never really spoke had only gone and become a doctor absorbing newfound confidence and resilience that has been a Godsend to me in the pandemic years. We all stood and recited the Hippocratic Oath (which has since been phased out/optional at graduation), and then the celebrations began.

Following qualification, you apply to train in your foundation years, and I was able to work locally to my home after years of long commutes. Two years of rotations through clinical placements, I cut my teeth as an FY1-2 doctor in adult medicine, A+E, general surgery, and psychiatry. The initial years made it clear to me I was destined to be a generalist not really settling my passion onto one area instead enjoying most aspects of all. Following this period, you then apply to your speciality of interest, and I was split between the adrenaline of Accident and Emergency but also swayed by the generalist and holistic appeal of becoming a general practitioner (GP). Due to the rota's and being a family man with young children at this stage I felt GP was for me but still kept my hand in at the local A+E doing regular locum shifts and being put on middle grade rota towards the end. Nothing could replace working in resuscitation bays and stabilising the critically ill using everything you ever learnt to bring life back.

I did my GP training in West Midlands and did placements in area's I had not experienced previously such as paediatrics, obstetrics, and gynaecology, completing a well-rounded training in 2014 when I became a fully-fledged GP. Working initially in my home locality; I loved the flexibility of locuming and spread my wings to experience a wider demographic and diverse types of communities in which to practice. I remained in the Midlands doing this until the scamdemic started where I worked in a large federation-based practice with big patient populations. This is where I will begin my story of covid 'dissent.'

Chapter 2 - The start of the 'apocalypse'

The preparations for Christmas 2019 were in full swing and bursting with tinsel, mince pies and Christmas presents, the news covered a cluster of cases being reported in Wuhan of a novel coronavirus that was responsible for a severe pneumonia in the Hubei province. I will be honest, after the first time I heard it, I filed it alongside other such cluster announcements of novel pathogens dotted around the world through my life and paid it little attention even as a doctor. However, venturing into the new year celebrations and more airtime was being given to this cluster becoming multiple clusters, initially of pneumonia without mortality.

As the month went on, claims of containment and limited human-to-human transmission developed into evidence of spread, initially to Thailand, with a real risk of international outbreak. The first death was announced on the 11th of January 2020, a man who frequented a local wet fish market, but digging deeper he had had a history of abdominal malignancy. The WHO seemed to have things in hand having set up an 'incident management support team' which had already sequenced the alleged virus on the 12th of January 2020 which it shared to the world with unprecedented speed. In March 2020 alarm was expressed by the WHO-China joint commission that 'deep concern' was being publicised.

On the back of this news came high caseloads on a cruise ship, 'Diamond Princess,' of 'asymptomatic' positive tests in addition to a secondary 'global hotspot' in Italy which led to an initial lockdown there, but a new concept of asymptomatic spread raised alarm bells for me. From then on, the spread of case clusters was shared with the world via mainstream media and despite a strict early 'no camera' policy being enacted in China, footage was being shared of sudden collapsing episodes in the streets of Italy and China and news of international clusters in the March and April of 2020 progressed. The tone and colour schemes of the media turned to red, black, and yellow reflecting the emergent threat of this alleged novel pathogen and eventually a pandemic was decreed on the 11th of March following on from the initial 'Public health emergency of international concern' announced by the WHO on 30th January.

I, like every other member of the public, aside from being a medic, watched on as concern levels raised and media coverage of the events racked up in length and intensity. Fear was contagious and anxiety and panic set in among family, friends, and colleagues during this unprecedented life event. However, this is when the confusion started for me as reassuringly, on the 19th March 2020, the UK government stepped in and officially downgraded the outbreak from that of a high consequence infectious disease to low consequence infectious disease (LCID) only to subsequently and inexplicably make the unprecedented step to lockdown the UK to 'flatten the curve' and 'save the NHS' as a one off event to stop spread and ease the burden of this LCID on the 23rd of March. This initial backtrack was a red flag to me as a doctor and made no sense, not only as it was an unprecedented measure but on the back of the downgrading it was ringing alarm bells. It was illogical and the collateral harm that such a lock down could have on the economy, mental health, NHS, and more, could not be underestimated.

Back in my day-to-day practice, I had worked at this surgery since October 2019, felt settled, enjoyed the team mentality, and wondered if I had found my forever home. As the pandemic gained pace in terms of fear and what I now reflect on as propaganda, the surgery was following early guidelines on handwashing and other standard infectious control measures as a precaution. In the February, the surgery reacted to escalating concerns by announcing via email that doctors and nurses that had any respiratory comorbidity were to 'opt out' of the frontline for their health protection and work from home. At this time, my experience was that things were very much quieter in terms of clinics being half to three quarters filled leaving a lot of time for administration, audit and professional development work in the interim time not seeing patients, so this didn't fit in terms of reflecting workload and witnessed morbidity. I was mortified as due to my asthma I would not be allowed to return to the surgery unless I signed a waiver to absolve the surgery of liability if I continued to work and get ill or die from Covid-19 illness. I spent all of five minutes seeing if this fitted with my ethos for patient care, and I simply could not have forgiven myself for opting out. I replied to say that I wished to continue and fight on the frontline despite the work from home offer and this was gratefully agreed to; I signed the waiver the next day.

I use the word 'gratefully' because as I entered the expected and ensuing 'furnace' of the front line of urgent and primary first contact care in the middle of a once in a life time 'apocalyptic' medical emergency, most of my senior colleagues were heading out of the building in the opposite direction to flee the arena and work from home leaving what I remember to be a very skeletal staffing of myself and a senior and respected 'old school GP' while some were literally dusting off old inhalers from yesteryear and distant past medical histories to absolve themselves from the front line. I lost a lot of respect that day for most of the team for doing the opposite of what, as a minium, you would expect of them. Huge respect for those in the profession, up and down the country, that remained to keep services safely ticking over.

Through the ensuing months and to summarise, workload was low, the propaganda of fear remained a constant thing, but things never really materialised. I clearly recall the issuing of our first set of PPEs for the brave few remainers. A single use blue surgical mask and single use white apron to be used all day, later superseded by vizors made by the local school. Wondering to myself if this was to be the next zombie apocalypse that we were being conditioned to expect then this was a very sad attempt at personal protective equipment to protect the frontliners especially when seeing the Hazmat's of the BBC coverage. Handshakes turned to elbow bumping, social distancing was a mainstay, the surgery decided to do its version of not quite a Tik Tok dance routine but a video promoting infection control measures and created a bit of focus due to low clinical workload. However, in the evening you would witness the fear turning up notch after notch, day after day, distinctly different to what I was experiencing in my days' work.

Prior to lockdown, we were given training in triaging alleged covid positive patients which essentially came down to 'can you breathe' or not. If you failed, the 'can you breathe test' we were initially trained in something called the Roth score which was a telephone triage attempt at guessing someone's oxygen levels. The crux was to get the person to count quickly from 1 to 30 in one breath to see where they could get to, the lower the number reached the lower the guesstimate oxygen levels, and these patients were advised to attend hospital for further investigation and possible oxygen supplementation. The issue was I never saw more than two people fail the Roth score in my time at this practice, both had severe COPD and would have failed this on a good day. Instead, I heard various incarnations of 'man flu' at worst, but on the whole a very benign symptomatology of self-limiting upper respiratory symptoms and malaise with a cough. Occasionally I encountered someone who couldn't smell or taste but that was an exception. I looked at what we should be recommending these patients in terms of symptom management, and no guidance was filtering through other than usual viral illness advice or rest, fluids, paracetamol, honey and lemon et al. Not quite the pandemic killer virus being reported on the BBC which, on those very evenings of quiet primary care sessions, the hazmat suit wearing, packed ITU footage had started to creep into the night's coverage.

Lockdown seemed to quieten things down even further, roads empty, ghost towns, people working from home, and a societal change that will be remembered forever built upon fear and propaganda spewed by the mainstream media. I was always concerned about the collateral damage from such measures; will elderly folk hold on to that chronic unintended weight loss through fear of adding to the workload of these already busy medics? (as reported by the MSM). What about the rectal bleed and chest pain not being felt to be worth risking going to the zombie apocalypse treating urgent care centre or hospital through fear of contracting the deadly disease of low consequence as per the narrative supplied by the UK government? At the same time, routine investigation and outpatients, as well as some routine surgeries, had started to be considered non-essential, again a grave concern for me of its collateral impact.

I was concerned at one point that I was getting the wrong tip of the iceberg working in the primary care and urgent care/out of hours setting where workload mirrored my earlier experience; quiet and low patient foot fall. At the time there was a local election and one of the candidates asked me if I wanted to take part in a public relations boost for her campaign and to help garner donations of food and snacks; I managed to corner the market in getting free chips, pizzas, and kebabs for the 'local hero's' at the hospital. Having considered the offer, I realised I could use this opportunity to see the other tip of the iceberg and observe the real frontline in action by donating my spoils to the local A+E, covid wards and ITU departments.

To summarise my experience of this campaign, I hardly recall entering any of the hospitals on any occasion of this month at the 'peak' of a worldwide pandemic where I was greeted by anything that I would call busy. I worked in the A + E department for many years and was known to the nursing staff, most of the time I would enter a very empty department to be swarmed for my pizza offering or free fish and chips. My suspicions were raised so I took my donations deeper into the hospital and to other hospitals to compare my impressions. Silence, empty wards, hardly a member of staff let alone patients. I once knocked the door of a so called covid ward to be greeted by a hungry staff nurse happy to take a pizza to share. When I casually enquired about the one patient, they had on the ward visible through the window, I was informed that he was in traction for a lower limb orthopaedic surgery issue and had apparently been presumed positive for covid-19 as he had coughed a couple of times during his induction to general anaesthetic. The ITU had no patients, a couple of standby nurses and an on-call doctor equally accepting my donation of a few slices of local pizza, the two dominos had been taken by the three nurses looking after the single man in traction mentioned earlier.

Independently of the ensuing scamdemic, my experience continued much the same, work was deadly quiet, but the BBC gave the counter narrative as did the 'next slide please' brigade of so-called experts showing an opposite increase in all things covid comorbidity and mortality related.

Due to family circumstances at the time, my wife and children had decided that we wanted to be by the sea and having toyed with the idea before and taking a number of investigatory working holidays eventually we took the plunge to move to sunny Cornwall for a life of sun, sea, surf, and work life balance, especially given the threat that the TV was warning us of, everyday life was put into perspective. By the time we sold the house we had made the move, and I started working in August 2020 to experience scamdemic part two.

Chapter 3 - Medical zombies

How has one of the most knowledgeable and honourable professions allowed this to happen I hear you cry? It has been extremely difficult to explain the lock step spread of ostrich syndrome running rampant with a culture of silence and cover up all too common in my NHS experience.

Part of this comes down to the selection of doctors that are essentially 'servants' to the pharmaceutical and medical institutions. Think about it, straight A students throughout, moulding their life experiences from primary school to high school with CV decorating hobbies and interests for that fateful medical school interview. Prior to recent years, medics tended to be ex private school students with a keen eye for toeing the line and academic excellence but from wealthy backgrounds where money is a strong motivator.

My insight into the difference between the learning methods engaged in the conventional medical programme compared to a conventional science degree, having done both, was stark in their contrast.

From day one in science degree, I was taught to question, hypothesis test, look for conflicts of interest, find root cause and proof for conclusion to form evidence-based research. Medicine was very different in that it was very much more teach by instruction. Lectures, seminars, encouraging rote learning, which is learning by memorisation of facts ad verbatim. The product of such training is the ability to regurgitate facts as told to you by the textbook or lecturer. Primed to follow rules, regulations, and guidelines, and to use their photographic memories to create algorithmic differential diagnoses and treatment plans focussed on medications at the conclusion of the diagnostic process.

Having been a doctor for seventeen years, and from mixing with many in practice, it is easy to see from this, looking back, the explanation of why the profession has been silent in its complicity in what I firmly believe will turn out to be the biggest medical catastrophe and criminally negligent clinical events in human history.

I always felt a lone wolf in my training and my subsequent career, the black sheep of the profession, never really having much in common or being comfortable around the culture purveyed in medical circles. Medics on the whole tend to be very OCD ironically, almost autistic traits by default. Alongside this is the pedestal upon which the general public place a doctor. This contributes to fuelling egotistical and narcissistic traits that are rampant across the board.

I recall my first day as an FY1 surgical doctor, I was assigned to work for a vascular surgeon whose reputation preceded him. He was very well known to be aggressive to his junior staff and had many complaints raised about this. Unbothered by his reign of terror and lack of remorse this continued a four-monthly cycle as his latest victims met him. I sheepishly walked into the theatre where he was due to operate on changeover day and as I scrubbed in to observe having not met him before, heard classical music playing in the background. As I walked in, the music seemed to increase in volume to drown out the bleeps of the machines keeping the patient asleep.

As my first step entered the room the music cut to silence and the surgeon stopped immediately what he was doing. Glaring at me, the silence was immediately intimidating. “Who do you think you are young man entering my theatre?” as if the scene had been set prior to my entry to humiliate. I introduced myself as his new FY1 and immediately he scowled. The music turned up slightly as if choreographed and the surgeon asked me if I liked classical music to which I said it wasn’t to my taste. The scalpel went down onto the ODP’s tray in disgust, and he stared fiercely at me. “How dare you not like classical music!” he exclaimed. If you do not tell me the composer of this piece you will never enter this theatre again. I made a blind stab at Beethoven, but it turned out to be Tchaikovsky. A grave error. I can’t recall the exact words he fired at me in disgust other than a rant filled with expletives that ended in my being asked to leave the theatre. For the next four months he barely acknowledged me, keeping true to his word I was never allowed to assist as I became his minion slave for the rest of my placement. The reason to tell this story is a glimpse into the mind, behaviour, and psychology of the 'old school' mentality and treatment of colleagues by the profession.

The very next placement I worked for a very strange fellow, polar opposite in his demeanour to my previous consultant. Very knowledgeable and quiet, three-piece suit for ward rounds and old school to a tee. The junior doctor day is usually 9-5 with a rota to cover what is called on-call for out of hours and emergency admissions. On a good day, 9-5 was a pipe dream, and I never used to leave routinely until 7pm as I always stayed around to speak to relatives to inform them of the day's decisions and condition of their loved ones. Not many people used to do this, a sad omission from the profession. Anyway, this doctor, irritatingly, used to run his ward rounds at 16.30 alternate days knowing that clock out time was thirty minutes later. He worked around the patients at a snail's pace, although meticulous in his assessments, he was always polite despite his habit of late ward rounds. By the time he was done it was visiting time, leaving a list of jobs for the juniors on the round to go off and do before the next day often finishing work as late as 9pm. This was an expectation not a choice. No one was ever thanked for their unpaid overtime, and no one ever got an explanation as to why he always chose this time to start the rounds. Doctors are very set in their ways and routines and again this contributes to the mindset of a medic and how and why things that fall outside of said routine or knowledge base make doctors very uncomfortable.

Expert patient's, aka those who have researched their illness, who ask questions and challenge the expertise of the doctor attending are not tolerated and bring about a silent aggression by the confronted doctor. The patient challenging is often marked down as a 'heart sink' patient which is an informal way of saying a patient the doctors dread seeing the name of. I remember a specific example of this in the early covid era, when an email was circulated about a certain 'anti-vaxxer,' a lady I had not met but in her 80's. The email warned staff of this patient and that she was, in a letter, pitching 'conspiracy theories' and 'misinformation.' They were further warned to be prepared for her if she booked in. I read the letter she had written and what was in front of me was a polite, knowledgeable, evidence-based rebuttal to certain measures, and queries around the speed and safety of the vaccine put together in an extremely concise manner. ANTI-VAXXER was the email heading. Pretty much this response to the 'reprobate' patient set the tone for what was to come.

Now I'm not saying all doctors have this level of narcissism or aggression but there is a culture in the NHS that is sinister in its universality. The irony is that whenever you start at a surgery or hospital trust one of the early presentations you hear is the introduction of what used to be called NHS whistleblower policy and how to report concerns latterly replaced ironically by the rebranded 'freedom to speak up guardianship' protocol. You are actively encouraged to speak out, voice concerns all in the name of patient safety and clinical governance. Until you dare to do exactly that. Be prepared to answer with your career and pick up your p45 on the way out. The fate of an NHS whistleblower is like a person meeting their end after a long stint on death row. The NHS has an endemic bullying culture and one of the victims of this culture is the whistleblower. I have myself raised many concerns focussed on unsafe practice or clinical incidents taking time to follow due process. I can't recall a single time where any affirmative action was taken on any of my complaints, which all ended up being swept under a very large NHS carpet. An abyss of unactioned complaints. The perpetrator who dared to raise concerns is then subject to a subtle, consistent process of bullying, and, the phrase is, 'managed out.'

As I have recently discovered, many of my online trolls, which I discuss in a later chapter, operate behind anonymous social media accounts, abusing, harassing, and stalking from behind anonymity. I have recently discovered the identity of many of them and I'm sure it wouldn't surprise you to hear that they include senior, highly decorated doctors operating abuse across social media. A novel way to oppress the whistleblower was born.

In 2025 the silence of the profession still astounds me particularly now that it is glaringly obvious that huge mistakes were made, and the profession ironically has been guilty of enormous amounts of false and premature safety and efficacy claims while in lock step denying any form of adverse event. Complicity is a driving factor here and alongside the other factors discussed in this chapter, imagine being a doctor who had administered thousands of these jabs to men, women, children, and pregnant women only for the realisation that they trusted you to be up to date and critical in thought enough to research and smell the rat that is clearly unsafe and defective. Heads are remaining down, and I am certain that the profession hope that it will all 'blow over' and be forgotten, but, I fear, the victims of disability, death and bereavement will not be so forgiving.

Chapter 4 - Mass formation psychosis

Before moving on to my experience in the Cornish Riviera and beyond, I wanted to share a few observations on the state of humanity's response to a life-threatening killer virus. Most of my friends and family, despite the fear campaign were fairly sensible initially but as panic and the psyops carefully coordinated by government agencies, such as the Nudge units, Spy-B et al as well as mainstream media, as soon as the mention of lockdown was announced in March 2021 something very bizarre happened to humanity.

We went from what, to me, was a quiet and tranquil period; people starting to find novel ways to entertain themselves, many people working in new ways from home, enjoying the fresh air etc., to insanity.

Things escalated quickly and a single lockdown to stop the spread became more prescriptive with limited time for exercise, 'essential travel' more draconian measures came in and as things progressed next came bubbles, 'traffic light' regions, rules of six, two-meter rule of social distancing and so on.

While it would be considered a normal fear response to stock up on a few emergency supplies and maybe the odd extra bottle of wine for good measure, overnight people developed a survival mode, mainly focussed on the toilet roll and flour/egg aisles of supermarkets. I will embarrassingly admit that given my work in the NHS I did take advantage of the early access to get my weekly shopping done and for that I am eternally repentant. However, even first thing in the morning the supermarkets shelves were low, a one-way system and traffic lights allowed people in and out, and there was a macabre eeriness to general life. While I likewise admit to putting an extra couple of items in my trolley for a rainy day and again embarrassing to admit the odd extra pack of loo rolls, I saw a new aggression in my fellow humans.

Gripped by conditioning of fear and prior experience of apocalyptic television dramas, people very quickly went into self-preservation and survival mode, this did not exclude doctors as we will see later in my diaries. I have witnessed grown men and women fighting over sixteen pack sizes of toilet rolls, the last bag of flour, putting a dozen boxes of eggs into a trolley with other customers entering disputes with them while removing them without the person agreeing.

The reason for mentioning this here is that we were seeing all the psychology from the government play out in one demonstrable example. Social media videos of the feuding loo roll brigade were shared leading to a ripple effect of 'whether I can maintain my perianal and genitourinary hygiene' throughout the lockdown and away from the ‘deadly killer disease,’ an odd reaction. I mention this just as a nod to the power of social media and why it was imperative for the controllers of the narrative to be aware of this tool and how this contributes, on top of the mainstream media (MSM) as a robust controller of public behaviour. This was exactly why, very early on, me included, there was a muted presence due to social media reaction at the sign of any questioning of the narrative.

Elsewhere too debate was silenced, including in coffee rooms at work. People were witnessing early cognitive dissonance, agreeing to illogical and overreaching rules, critical thinking left the building as people entered a state of inconsistent thoughts, beliefs specific to behaviour and attitudes to themselves and others. We were also witnessing a mass formation psychosis in real time whereby large groups of people experiencing similar circumstances develop a group think leading to a behaviour change but reacting in an exaggerated manner to a perceived threat. The relevance of this short chapter is to add some context to the reaction of the medical profession to my own covid dissent.

Finally, just at this time, sceptics of the events were starting to ask questions. It was social media that led in showcasing the covid measure dissidents being chased around by police for spending an extra five minutes over the allotted hour running on the beach. The public being pinned down to be forced to wear masks, friends having a coffee on a park bench for a few minutes too long and sitting too close, as well as rumbles of discontent in the form of organised protest being extinguished by baton brandishing police manhandling the owners of dissident voices. Again, all contributing to a mass formation psychosis but perpetrated by those tasked with protecting us. This has a direct correlation to the NHS subsequent response to pandemic management and response documented throughout the rest of this book. This chapter also heralds the incoming herd mentality and group think led by all involved in the illusion of the scamdemic. When influenza disappeared from the earth to be replaced by the novel coronavirus diagnosed by… well, any and every symptom including none. A deadly pandemic so bad that you had to eventually test for it to see if you had it as you will see in the ensuing chapters. Where, like a deadly assassin, you could kill your granny with a disease you had no signs or symptoms of having!

Chapter 5 - My Cornish dream

I remember the moment of walking into my new dream Cornish home, initially alone, as I travelled ahead of family with the pooches. It felt like a new start. Away from the the hustle and bustle of towns and cities of the midlands. A life of sea air, sandy beaches, coastal walks and hopefully surfing and water sports. My long-held dream had come true as I set sail on a three-day working week and four days to enjoy my family, friends, and new surroundings.

I had worked for the surgery that I was initially employed by on my working holiday's and was extremely happy with my new home. My initial stint at the practice was uneventful other than the usual rigours of daily practice and the workload and time constraints of the job. COVID impacted things very little other than the testing that had begun in April 2020, and 'social distancing,' as well as a plethora of inconsistent and medically illogical measures such as mask wearing while out for a drink yet the protective benefit of eating a meal meant that you were exempt from wearing a mask.

The dawn of testing came on the background of quiet covid times both in my practice and in the media, things had seemed to calm down. I certainly hadn't seen waves of covid mortality and morbidity in my months prior to testing being introduced. Boxes and boxes of the hallowed lateral flow test were delivered (LFT). At a similar juncture PCR testing had started to rack up in popularity and confirmed many 'cases' (discussed later) of people who were well, in an attempt to break the 'chain of transmission,' and car parks converted to mass test centres.

I clearly remember the comments of my colleagues stating the inadequacies of the LFT's and the more inquisitive among staff members who understood the technology behind PCR stating low specificity and sensitivity, our manager at the time proclaiming the tests not being worth the paper/plastic they were written on!

As the fanfare died down, most of the medical profession in the light of flawed PCR and false positive LFT's reacted to the incessant waves of the 'testdemic,' again I contacted many of the so called patients to see how they were feeling. Only to conclude they had succumbed to a self-limiting coryzal illness not usually put through to be discussed with a nurse or doctors by the reception staff.

From initial doubt and with a critical mind, given the influence of the mainstream media, initial concerns were parked, but I watched medical staff fall hook, line and sinker for asymptomatic carriage and tests leading to define cases. Deaths were reported in a very odd way in that a covid death simply was someone who died 'within 28 days of a positive covid test,' an unusual metric to document pandemic mortality. You could have died in a car accident ten days after a jab or died of a heart attack in hospital (where tests were done at triage) and incidentally be covid positive only to succumb to the primary illness or injury but be included in the covid mortality ticker tape running across screens worldwide.

One of the things I noted was speculation about how the world might exit from this terrible asymptomatic illness… the covid 19 'vaccination.' All eyes were on this momentous occasion when 'Maggie' a 91-year-old lady stepped forward on 8th December 2020 to take the first ever dose of the worldwide circuit breaker for covid death and morbidity, the novel gene-based injection marketed under the title of vaccine despite the reality of its novel gene-based mode of action i.e. not a vaccine.

I had a short period in the interim, working for another local surgery, my decision to leave was due to measures outside of pandemic concerns, mainly to do with rising workload and, in my opinion, shorter patient encounters compromising patient safety. Between April 2021 and September 2021, I worked for a short but enlightening period at another local federation practice across numerous sites.

It was here that I had encountered a small number of early but inconclusive possible covid jab reactions. The fanfare of the first delivery of covid jabs to a hero's welcome followed shortly behind by the liquid nitrogen ice clouds given the initial need (soon forgotten) to keep the contents super chilled and injected within a short period of time after delivery. All hands to the pump and it was my one and only shift working at the jabbatoire where checking people in, that I witnessed the practicalities first hand of this mass vaccination programme. Let it be placed on record at this juncture that I have never personally recommended or administered a covid jab in my whole career.

People entered the building in their droves, asked minimal questions to check eligibility and inclusion/exclusion criteria, and were formally added to the massive queue, stopwatch in hand to ensure their ten minutes of non-instant death/reaction from the jab. There was literally no information exchanged between injector and patient hence my concern of how informed consent could be achieved. After about a month or so, the safeguarding of the stopwatch observation post administration swiftly stopped as practices rejoiced as more punters could be moved more quickly through the jab process and of course, the more people who could be jabbed, the more money could be made with some doctors boasting how many thousands of people they hit with the shot… at around £12 per jab to the practice.

One of my immediate concerns at this time was regarding the temperature at which the novel gene-based jabs could be administered. Initially there was a strict cold chain and use within 60 minutes. This soon disappeared from the protocol. No mixing of brands for subsequent jabs was swiftly superseded by any jab they had in stock. After an initial spell of ironic jab snobbery around Pfizer being perceived to be better than its competitors, Moderna and AZ (now discontinued), people were receiving various mix and match combinations of brands that had simply not been tested as combined treatments in clinical trials.

I spent a time at this surgery shortly after the jab roll out. I swiftly encountered a few minor, but many major, reactions to the jab. These were not exclusive to brand at this point with people reporting odd reactions of prolonged malaise through to reactive arthritis and some other 'rare' initial side effects not limited to neurology, but many clot related, inflammatory and cardiological events in previously healthy young people. Ironically, demographics that were not at risk of the illness were experiencing side effects exclusive to the time period following their covid-19 jabs. Now I agree rare coincidences happen, but if 'rare' keeps happening and 'coincidences' keep occurring, this raises the flag that they are not so. This was my experience and as we all witnessed, the rare and coincidental simply increased in line with the magnitude of the roll out.

What was my eureka moment that led to my full commitment to questioning the narrative? The honest answer is a two-week period, which ensued a small number of months after the roll out at my new surgery. It started on week one where I received a screen note that I was to make a condolence call to a patient's husband who had lost his wife; a mother who had died from what had clearly been documented as a vaccine induced thrombocytopenia and thrombosis (VITT) death following an Astra Zeneca (AZ) jab. This call sickened me to the core given that I am also a father of four children. A matter of days later the surgery received a further death certificate, a female patient two years older, again a mother who had coroner confirmed death as VITT as per previous patient discussed. This was not a coincidence given a small uptick in DVT's and PE's being documented at the time of AZ role out.

It was the week of the second death that I was working as duty doctor and received a call from my local neurosurgical colleague from Derriford hospital who told me of a forty-one-year-old male who had collapsed in a nearby town, seized and found to have a cerebral venous thrombosis (CVT). I noted this young healthy male had received an AZ jab a day prior and I wondered if this had any role in his CVT. This was the first time I had ever been ridiculed and questioned by a colleague about covid issues and despite a temporal relationship to the jab and clot and in context to the previously mentioned deaths of two young healthy patients. He asked if I was 'one of those anti-vaxxers.' Charming.

I was alarmed and upset by this interaction and being dismissed as an 'anti-vax' crank and approached my senior ex-military colleague who showed zero concern particularly the brand correlation to AZ, instead advising me disturbingly that the 'show must go on.' That very weekend the surgery went on to jab many folk from my local community including with the AZ shot that was raising huge concerns with me based on recent events; concerns that I was not alone in having, as time would tell.

At this point, I resigned for a number of reasons not least I felt that the care fell short of acceptable standards. Ironically, I went back to my original surgery for a second attempt.

My time at this surgery initially went without drama or any event of note and I had settled back into normality as a GP in rural coastal practice. However, it soon became apparent that things had started to go awry in the vaccinated community.

My initial concerns came from a short audit I performed where I noted that those suffering from 'covid' (well at least 'testing positive') on the whole were the vaccinated which made no sense and provided a paradox (discussed in following chapters). What became immediately apparent was the number of patients experiencing ill health across the spectrum of not 'feeling ok after the jab' in general terms, to full blown novel auto-immunity, reactive arthritis, young folks (as the age of vaccination reduced quickly dropped - a further concern) suffering from neurological sequela and cardiac/blood clot issues. This was the last straw.

The audit I did was presented to my colleagues demonstrating covid positive PCR from labs was mainly affecting the jabbed patients and unvaccinated patients represented sparse positive cases in my small review. I also made my senior colleagues aware of the adverse events I had experienced at this and the previous surgery. All concerns fell on deaf ears under the mass formation 'safe and effective' psychosis. Despite my in house attempts to express concerns, the jab roll out continued full steam ahead, seemingly ignoring mine and a small number of tentative early roll out fears locally. A local nursing home had a sharp uptick in venous clots shortly after the jab was administered with nearly a quarter of residents having a clot in their chest or leg within a short period of receiving the jab. A concerning and noticeable red flag waving for all to see but strangely ignored.

I tried a more academic approach to raising the alarm and decided to write a short presentation including data on early concerns regarding clots and myocarditis in the young from published peer reviewed journal articles and studies. I offered to present my evidence-based audit and literature review concerns (a slide set yet to ever be presented) but was faced with resistance. I offered to make the presentation at a local meeting of surgeries. My offer was declined. This was the first early sign of ostrich syndrome among my colleagues. I remember clearly taking a journal paper into our coffee room and as soon as I started to speak, journal in hand, a half dozen of my colleagues stood up and walked out, one proclaiming that the break was their down time and they didn't want to 'listen to my conspiracy theories.'

At this juncture I decided to resign yet again from my post, despite job satisfaction and a good work life balance, on the principle of ethics and in particular the roll out targeting younger healthy cohorts notwithstanding covid, as the narrative went, being indicated to only be lethal to the elderly and comorbid. They also launched a campaign to jab 12-17-year-olds and pregnant women. I was extremely angry to see this happening and it being accepted by colleagues without so much as an ounce of critical thought. The main driver to the resignation was a private conversation in the coffee room among trusted colleagues where two senior doctors, two senior nurses, and an ANP, declared they agreed that this was an unnecessary measure and one they would not be offering to their own children. This was backed by a collective statement that no one of that age demographic would be subject to covid-19 jabs 'under our roof'… so no jab for their own children, however, the VERY NEXT DAY the surgery posted an advertisement for the jab centre (what I referred to as the jabbatoire - a centrally located mass 'vaccination centre' in my hometown) to take your children for their first covid-19 jab under the premise of safe and effective. Aside from a lack of clinical indication, that those administering it were so sceptical that they would not give it to their own children, speaks volumes.

For me, this was the last straw. After trying to raise concerns in the correct manner to the right people only to have my observations fall on deaf ears and knowing that affirmative, sacrificial action was required, as per my Hippocratic oath and as per GMC professional standards. I had to make a stand and speak out. In effect, the GMC, by contravening their own guidelines, forced me to launch my social media career in earnest as I sought to blow the whistle to do all I could to protect the public.

To be clear, I was never one for social media but decided this was the best way to reach the masses. So, I put pen to paper or rather pixel to screen, and wrote an essay as to how, why, and what my reasons were for making my dramatic resignation. I presented it in layman's terms and laid out my safety, ethical, and evidence-based concerns. Things quickly went viral, and I was asked to appear on lots of podcasts to give my views, which were never allowed to be aired on mainstream media. The public and regulatory scrutiny of my own practice stepped up shortly after this in addition to military grade social media censorship which I discuss later.

I will now post this career ending essay in its entirety over the next few chapters, for you to read and assess its integrity and morality in its published version which was subsequently well supported and distributed on many platforms across the social media and podcasting community leading to some very unwanted collateral attention. This was written in February 2022 so please forgive me for my many misjudgements in the essay; as they say hindsight is a great thing.

Chapter 6 - Breaking the silence - the beginning of the end!

My name is Dr Dave Cartland. I am 39 and a practicing GP in the UK, down in sunny Cornwall. I qualified in 2014, and prior to that worked my foundation years in the West Midland's Foundation programme and GPVTS (Black country).

I qualified from Birmingham Medical school in 2008 as a graduate entry medic, having completed a Biomedical science degree in 2004 with a large component of this reading in immunology, virology, microbiology, and medical statistics.

I proudly qualified with first class honours and went on to publish work in Angiogenesis as part of the Birmingham angiogenesis research group in 2005. I am a married father of four awesome children who are my life and have two faiths… Jesus Christ and Aston Villa FC.

Anyway, enough about me. Why did I become a Dr I hear you cry? I became a doctor to help my patients, to be their advocate, to help them in their biology, psychology, and social circumstances.

I will always remember exactly the moment of my graduation when we recited the Hippocratic oath. Part of this powerful oath is a vow. A vow to 'Primum non nocere'- first do no harm.

I hope my patients would agree that I am a caring, decent GP. I enjoy my job and the role I have to play in patients' lives and can safely say this vow has formed the basis of my medical career thus far.

To not recognise, notify or publicise concerns of harm would be contrary to mine and my colleagues' oath taken at qualification. I am writing this as a commentary and as a personal reflective piece, some of it is opinion from anecdotes, others statistical or government data, I have no objection to it being shared.

When the COVID pandemic hit the UK, the confusion, fear, and medical uncertainty was palpable by colleagues and patients alike. I want to say from the outset that I am by no means a covid-denier, I have seen a few people debilitated acutely and chronically with this disease albeit it rare and in the elderly/comorbid.

I rolled up my sleeve to take the jab despite reservations about the speed of the development and lack of prospective safety data but to 'do my bit' in the face of unknown aspects of a pandemic. What sort of person let alone Doctor would I be if I didn't step forward to help 'protect the vulnerable' and of course my own health. I took my first Pfizer jab on 13/1/21 in complete good faith of the science and public health underpinning its roll out, trusting, as all patients do, the integrity and reassurance of the government and its scientific advisors.

Six days prior to that, on the 7th of January, I lost my best friend to the 'pandemic' a driving factor in the decision to take the jab.

At this time, only very short-term vaccine safety data was starting to filter through and obviously medium - and long-term safety data for the Jab was absent, but with the clear and present danger of the disease, a clear weight in the balance for taking it. All seemed logical to this point.

In my capacity as a GP, something that struck me from the very beginning was that (thank God) the number of dead, dying or seriously unwell patients never hit the heights of what I was expecting or that was being projected from the various modelling data that was being shared at the time.

'Nightingale hospital-gate' as I will call it never took off, and mortality wasn't as high as the fear levels had suggested which was great news. This wasn't just an independent observation. Now as previously mentioned ITU's and A+Es were allegedly full of covid-related morbidity and mortality, however as time went on, slowly but surely, some things started to seem odd and not make sense to my pre-pandemic experience of medical practices.

Firstly, and very early on, death certification changed, and all the safeguarding related to a second 'part 2' Doctor verifying events around death to come to agreement about what we call the medical certificate of cause of death (MCCD), was over-ruled.

This was reasonable on the face of it given the swathes of deaths that were expected, some of the 'paperwork' would understandably need streamlining. Over time I started to see death certificates coming through in patients that were generally severely unwell with multiple comorbidities as MCCD 1 (a) COVID-19, the main cause of death felt by the doctor or coroner.

When looking into specifics and at the notes I was seeing when they had been admitted to hospital, they had clearly 'entered the building' having been unwell with other things: cancer related issues, end stage COPD exacerbations or renal failure, broken bones, strokes, heart issues, a barn door case of urosepsis died in resus.

On entry to the mortuary marked down as 1 a) COVID 19. Out of hospital deaths followed suit. All it seemed to take was the mere mention of 'cough' by a relative elicited by the Part 1 Doctor (a common symptom around the time of anyone's death) the MCCD would be confirmed... COVID-19! This was happening with alarming regularity. Something odd was happening and didn't sit right.

The next irregularity seemed to be around testing and the early use of statistics. I won't mention too much detail here but the testing for COVID 19 was using PCR (the results of which have formed the mainstay of covid surveillance and shielding protocols throughout). With some prior knowledge of this from my science background and an early statement from the inventor of PCR, Kary Mullis, stating that this was not a technology purposed for diagnostics of viral infections.

To use such technology, in knowledge of this, to gather 'cases' just seemed very unscientific. PCR is used to amplify a small sample of genetic material and to increase its 'visibility,' in order to genetically sequence that small initial sample with common practical application in things like forensics.

We had technology to isolate the virus, electron microscopy among other things but not a mention of it. It is commonly accepted that when using PCR, you have to set a cycle number of amplification. The agreed limit to get a good amplification and reducing false positive 'signal' is 20-25 depending which data you look at. To go over and above this cycle threshold yields very high percentage of false detection.

Multiple FOI requests have been returned from many of the testing labs stating that 40-45 cycles have been used as standard throughout the pandemic. Consider this in simple terms as a game of Chinese whispers, in the amplification process translation errors commonly occur in decoding and an early error will pass through to the next and subsequent cycles.

Just like Chinese whispers the initial message after 45 people may be very different to the original word in the game/PCR sample compared to 25. Again, warning signals. Essentially if you turn the cycle number up high enough everyone will have a positive result whether true or false positive. This again defied basic science but was still used despite these very well-known limitations.

Furthermore, the medical statistics being used were and are still vague such as 'death within 28 days of a positive covid test,' about as nebulous as you can get for such key figures important in a worldwide pandemic seemed to be being purposefully blurred.

Let me expand here briefly, surely just report people who have a case definition of what the viruses' physical symptoms were (again ever changing) and a positive test = a case. Someone feverish with cough and short of breath with a covid positive test prior to their passing was of course likely a death FROM COVID-19; it wasn't seemingly that 'simple.' It certainly didn't seem good medicine to report a person in intensive care for example who died of a PE secondary to his elective total hip replacement and passing away from this as COVID 19 due to asymptomatic positive screening, this just didn't add up!

For medics even now simplified data is hard to find, with guidance ever changing almost weekly and finding straightforward data e.g., how many of those 150,000 recently surpassed deaths were actually FROM… not WITH COVID-19 seem impossible to lay hands on. Surely for such an inherently important statistic, clarity should be paramount particularly in order to counsel our patients and risk assess for ourselves as independent practitioners.

Again, recent FOI reports I am being made aware of state in one example only 2 of the 97 reported covid deaths fulfilled the disease definition and positive test criteria among others; alarming stuff!

Back to personal experience, the time came for Jab 2 with the 'friendly warning' that mandate was coming for NHS staff and after careful consideration and using evidence-based approach, myself and my wife reluctantly agreed to partake; my career weighing heavy on mine and my family's mind. I will say that at that time the data wasn't as black and white as prior to Jab 1 but for the greater good and all that! Meanwhile carers were being unceremoniously sacked showing the governments clear intent!

Questions on mine and my patients' minds were… Why do I need this again so soon? Does this not show the jab isn't working? Jab one made me quite unwell for a number of days with some very strange symptoms what if this happens again? What if jab two doesn't stem the tide, surely not a third? etc. Just around that time, a few early signals of a wide selection of 'Jab reactions' were starting to find their way to my attention through patient contacts… more on this later.

Chapter 7 - What makes a case?

I briefly touched on this above but want to expand. It sounds elementary to say this but to have a medical condition or to be called a 'case' one must satisfy a basic criterion… you must have symptoms and the presence/evidence of the disease such as a scan or a test.

For example: dysuria and frequency plus E. coli in the urine = UTI (see Koch's postulates), headache/neurology plus cerebral mass = brain cancer, abdominal swelling and baby in-utero is pregnant… You get the gist.

However, in the case of COVID-19, people were being told something I have never heard before in my career. You can be a 'case,' a dangerous 'threat to your gran,' without even having a symptom. A dangerous spreader of the virus without so much as a sneeze. The hallowed 'asymptomatic carriage' (AC).

From my basic virology knowledge and in view of alarming data re false positives as above, all was a little confusing. Asymptomatic disease is true in other areas of biology for example bacteria and some protozoa can be asymptomatic. Sorry to freak you out but your hands are probably covered in Staph Aureus (a natural 'commensal' bacteria that colonises skin), and likely coliforms (a medical word for bowel organisms)- despite how much hand gel you are using. They are single independently alive unicellular organisms that reproduce asexually.

Viruses on the other hand are a little more complex. By definition, they are obligate intracellular parasites. They cannot live very long at all outside of a human cell as they are fully dependent on human cellular machinery (nucleus, ribosomes, and various enzymes) that they borrow to allow them to reproduce.

A single virus enters the human cell, takes over the genetic machinery, and causes programmed cell death 'apoptosis' of the cell. Having ruptured the host cell and bursting with an exponential number of virion's compared to the initial individual invading virus those new viruses go on to then infect a multitude of neighbouring cells and before you know it innumerable viruses.

Cell damage is left in the wake of this and the consequent immune response to fight this vicious attack. Part of what makes you feel unwell from having a virus is the immune system response aside from the effect of cell damage by the virus. Cytokines, inflammation, and pyrogens storm the body to gain control before a critical mass of virus can take hold. As only males of the world know this is particularly present in the seasonal illness known as man-flu, the consequent myalgia, rigors, nausea, malaise, and near-death experience of man-flu is in part due to this.

I will summarise it this way… and apologies for the lecture

on virology… if you have a virus particularly a 'deadly' SARS type coronavirus respiratory tract virus, you KNOW you have a virus whether it's a sniffle, aches, pain but still not asymptomatic carriage especially at pandemic level.

Incidentally, the way pandemics become endemic is by exactly the mechanism we see being displayed in Omicron data. As previously mentioned, viruses are parasites and have no interest in killing their host. That wouldn't make much sense from a survival plan perspective.

Viruses learn to become symbiotic with their host. The perfect storm… spread like wildfire/high transmission and low pathogenicity/harm to the host, survival of species. The misuse of this medical 'phenomenon' of AC (Affirming the Consequent) and not acknowledging the fact of what is happening now with so-called Omicron being a much milder disease, is in my medical opinion misleading the public.

A virus pandemic will always become endemic, fact. The AC fallacy use has driven fear into our society as we mask up, avoid others, keep our two metres, and decide not to visit our elderly relatives is implausible. Aside from that, the collateral harm from all the psychological, physical, and social/economic harm is a ludicrous idea in the face of such a lie.

To 'LOCK DOWN' is inconceivable and unprecedented as evidenced by having never been used as a management strategy for pandemics ever in humanity.

Talking of collateral damage and just while we are on the subject of harm from such an erroneous medical 'fact' and subsequent policy. When does the harm to patients for such damage become an important denominator? From fear of seeing the GP, delayed cancer diagnosis, bottlenecking of NHS services/inflation of waiting lists, palpable mental health distress, social issues such as poverty and loneliness, the list is endless. Surely to allow all of this as a trade-off, the threat needs to be 'bubonically' high, and the risk of treatment low, to allow such measures to become acceptable.

Silence is deafening

Over the last few months something has become startlingly apparent in regard to the latest data from Omicron which didn't seem to correlate with previous waves. I say that from the outset, hence my discomfort with current policy in particular the mandating of a medical treatment for staff.

It came about after I was asked to consider the booster (in English a third dose in six months of something I had twice in the recent past) that gave me side effects and something I and probably no other Doctor can say they have experienced before in their medical practice. A tri-6-monthly jab with speculation of even more - Flu Jab annually, Hep B jab lifelong, the ten yearly tetanus et al - something didn't feel right. To take one (again) for the team or not?

The answer to this dilemma was simple. The 'vaccine' needed to confer both personal benefits i.e., if I take it my risk of being ill and even death would/should be reduced, likewise in the interest of being a good citizen and medic of the world that it would also reduce my ability to catch and likewise transmit to my relatives and patients (aside from evolving safety data which I will discuss later).

I took a short look at the data from my own surgery (admittedly low numbers, to try to spot a trend). We receive positive results as a batch each week and one of our ANP's (Advanced Nursing Practitioners) contacts them all to see how they are. At this point I will say that most of the positive results displayed mild flu like coryzal symptoms, evidence for me that we were dealing with a different beast in Omicron which was really great news.

I counted 102 'cases' of positives in the first two weeks results and traced their vaccine status: positive… treble vaccinated.

A pattern was apparent; long story short in that two weeks 94.1 percent of the patients n= 96 of 102 were treble or double jabbed (mostly treble) and just the 2% n=2 unvaccinated.

I decided to repeat this the next week and this time 100% of the 38 patients were double or treble jabbed zero unvaccinated. This seemed odd and quite contrary to what I expected but it had put some data to the gut feeling of what I was hearing of in clinic.

I decided at this time to see what was happening nationally using UK and Scotland gov.uk surveillance figures and was startled to find similar findings.

In one study initially looking at just positive cases in UK an over 100,000 dataset showed 89.7% positive results over a three-week period treble vaccinated v 3.7% unvaccinated. Even in the knowledge that the lower proportion of unvaccinated as a percentage of total population these percentages were too far apart.

A recent Scottish data set (weeks 1 and 2 of 2022) despite showing reduced outcomes in three different outcome categories, the ratio of unvaccinated, 28%, to vaccinated, 72% of the total Scottish population. Roughly speaking, outcomes of the parameters of 1) positive test 2) admission to hospital and 3) death was seen to be represented by 80% of the total numbers in treble vaccinated status whereas only 20% were unvaccinated in proportion.

Why were people vaccinated for a disease then go on to die in a ratio of 4:1 from the disease they were trebly protected from? Alarm bells ringing… again.

At this point, I committed time to fully researching whether the risks of a jab were proportionate to the risk of the disease. Networking with other medical professionals constrained by fear to not report observations or whistle blow, the sharing of papers, research data and individuals' clinical concerns and anecdotal observations were forming a clear pattern and hypothesis.

Study after study, data set after data set, seemed to conclude that the vaccinated group appeared to be at higher risks of catching covid despite full vaccination status and in the above data, higher admission rates and death.

The decision was becoming clearer as to whether to take jab #3. At this point I concluded the Jab wasn't in any way reducing my chances of catching or spreading the virus, in some cases increasing the risk (will avoid commenting on Antibody Dependent Enhancement- ADE in this article here but coherent with this principle).

I was at this time being told that nasal carriage of the virus was much higher in the unvaccinated, but again this wasn't bearing out in the cohort studies that I was seeing published. So, my conclusion was the Jab wasn't (in my humble opinion) in anyone else's interest so that only left the conference of personal safety from becoming seriously ill and its obvious major repercussions when it comes down to rationale to mandate a medical treatment.

This will be visited later when I briefly discuss medical ethics around consent and lack thereof.

Chapter 8 - Signals of harm

Alongside phenomena discussed above, I was starting to see flickers of what I am calling 'signals of harm.' At this time, I felt slightly isolated professionally as no one would seemingly enter deep debate on the subject, offer alternative explanations for data, evidence of safety, all at the same time as hearing from the government that the vaccine was 'absolutely safe' in children and pregnant women under the guise of adverts with the headline 'the best way to protect yourself from Omicron is to be treble vaccinated.'

This didn't seem true to me.

Bit more virology while we're here… I saw an advert that could only be described as propaganda and misinformation on the TV where a family were all mixing inside the house and small black particles of virus were pouring passively from all of the family members mouths into each other's faces in a bid to promote space and window opening.

I couldn't believe my eyes.

Respiratory viruses such as SARS-COV 19 are spread not by asymptomatic carriers as previously covered. I may be speculating here but I personally feel asymptomatic 'carriers' are mislabelled false positives as previously discussed. Respiratory tract viruses are known to spread by aerosol generation i.e., 'coughs and sneezes spread diseases.'

None of this was science as I know it! It was 'The Science' of propaganda.

I just was not seeing the degree of harm from the disease borne out in gov.uk covid surveillance data despite high case numbers and certainly had not had sight of safety data to support such a statement to the degree of justifying jabbing five to eleven-year-old children and pregnant women and their unborn babies.

We all know as clinicians that prescribing in pregnancy is riddled with lack of safety data and manufacturers often sit on the fence with recommendations of erring on the side of caution and avoid prescribing unless benefits far outweigh risks.

To accept a medical treatment of any description you have to weigh the benefit v risk of the treatment v the disease. I recalled at this time at the beginning of the pandemic that they were showing personal stories of individual patients who had succumbed to the disease and remember saying that a lot of these folk seem to be generally overweight.

Recent plethora of data sets show that multiple comorbidities go hand in hand with high-risk of death or serious adverse outcome in addition to age and immunosuppression. Despite rolling out the Jab to anyone in the world over five this didn't seem to consider the individual's personal risk, i.e., the smoking eighty-year-old diabetic vs the fit, well, twenty-year-old sportsman, have completely different risk profiles for becoming unwell from COVID. Why was this not being extrapolated; we were jabbing people with a treatment they simply DO NOT NEED or at the very least offering negligible benefit as you went up through the age groups?

Chapter 9 - Do no harm

Initially the flicker of harm signal subsequently became a flame and most recently a blazing fire!

My individual experience of seeing very odd post-vaccination reactions came ironically around the time of my second jab. In just one week, I saw a terrible case of a very fit gentleman who was suddenly unable to move, his hands and feet and became swollen, a reactive arthritis, coming on inexplicably without any prior rheumatological history, two days after his vaccine.

A couple I did a home visit to came out with a most bizarre skin rash with large ulcers appearing widespread to the body the like of which I have never seen, again within a week of the jab and with no prior history of skin problems.

While duty Doctor I recall a bizarre conversation with a medical registrar who advised me that a patient needed to commence anticoagulation for a clot on his brain which was triggered by his sinus inflammation seen on the MRA scan… (The gentleman advised me he had never so much as sneezed in his life); he had received his jab four to five days prior.

And then the worst. Two forty-year-old female patients within about a week of each other and each within close proximity of the jab (1-2 weeks). Both died without significant medical history. I will always remember them both but one had two young children of similar age to my own, MCCD VITT (vaccine induced immune thrombotic thrombocytopenia), catastrophic clotting to multiple systems leading to death.

You may be surprised to hear that the coroner did attribute the outcome to the jab. A rare phenomenon.

Day after day I was seeing on social media comments from friends' family and strangers making causal accusations towards the jab of a multitude of harms. Some serious harms too including new and permanent/progressive neurology/fits/collapses, dermatological presentations, menstrual changes, immunological, and simply not feeling 'right,' the list went on and on.

This was just the start. I joined a group called NHS 100k in the wake of opposing the upcoming vaccine mandate and was again hit with a barrage of harmful episodes within close proximity to the event of a COVID jab.

This was seeming to be more than just coincidence purely by noted incidences.

I was reading and hearing on official media that reactions were extremely rare, but I was wondering how rare does rare have to be for it not to become common? Everyone seemed to know someone who had fallen foul. I looked on the yellow card reporting system and VAERs but the incidence of what I was seeing from experience didn't match the data from these resources and seemed understated.

I saw an article today from MHRA data which showed huge increases in reaction to the covid jab across all manufacturers at unprecedented scale compared to any other previous Jab ever and consistent across all manufacturers.

I eventually wrote a post asking for people to contact me personally on social media and joined a telegram group 'They say it's rare' (they also have a website). Scary amounts of people stepping forward with a full range of alleged reactions in proximity of a jab.

In the last 48h I have personally had over 200 DMs with personal stories of post-vaccine injury reaction and even deaths and the aforementioned group is just constantly posting these stories.

Going back to my signal of harm, there was an alarm sounding, and as per me writing this article, once again I was reminded of my Hippocratic Oath. The above may just be coincidental I am well aware, but my oath was to do no harm (reminder at this point: I have never jabbed a patient with the COVID vaccine).

Aside from this, more worrying observations from different arenas. As an avid football and sports fan, reports of footballers and fans dropping like flies, some live on TV, and numerous tennis stars falling ill during games (whether vaccinated or not, reports are sketchy). Pilots dying in never-before-seen numbers. USA reported an increase in life insurance claims by up to 40% in the previous three months. All-cause mortality in certain periods of time higher than ever before. Funeral directors and doctors going public on unprecedented numbers of young folks 'dropping dead' and post-mortems of VTE causation. Increased numbers of referrals for uterine cancer… the list goes on.

And not a word from the mainstream media.

To claim anything negative is shutdown, Orwellian 1984 style. Nowhere on MSM is this potential for causality even whispered. If it's all coincidence, then where is the debate, where are the scientists calling it out and debunking such 'conspiracy theories.' There doesn't seem to be any correlation to the footballer phenomenon other than a couple of ex pro's stepping out on their private social media platforms.

Censorship gone crazy

By now it has become clear, one mantra, one narrative: After locking us down, then paying for half of our meals, then discussing herd immunity followed by the rule of 5/2 metres/re-lockdown (nothing like consistency) they eventually found consensus.

Vaccine, Vaccine, Vaccine.

The lack of discussion of potential harm suppressed alternative treatments with some solid medical grounding data e.g., Zinc/Ivermectin/Vitamin D, with no apparent push to offer general advice on this as 'COVID prophylaxis' if you like.

To debate and seek evidence-based medicine or to offer an alternative viewpoint or to even speak freely, a right that was fought hard for, seems to have succumbed to censorship. For example, who better to speak about the mRNA technology being used than Dr Robert Malone the very experienced and humble scientist who had a huge role in its research and application in vaccine technology.

Not debated, simply deleted, platform removed for a concern from the 'inventor' that there are risks associated (to hugely understate his viewpoint) is a view I would clearly like to give credibility and time to and would wish to hear his voice.

Dr Mike Yeadon, Dr Peter McCullough, and so many other eminent and qualified doctors and scientists, all vastly qualified, published, and relevant to the covid debate given their credentials, limited to back street media platforms and underground presentation of their opinion in opposition of the major narrative.

This is unprecedented in my career!

I have fallen foul of the Facebook police on a number of occasions, having been sent a recent article by a colleague- with a very sensationalist headline- 'Covid Vaccine Scientific Proof Lethal' (Jan 2022), but aside from that it was essentially an article cataloguing published literature of case studies, small cohort studies, prospective studies showing evidence of harm, post-mortems and the like showing causal links to potential harm.

Ninety-day warning from FB for the crime of one of the articles stating that the jab is being given under the emergency use authorisation. This is a fact that I fairly-well recognised to be TRUE… article removed no ability to appeal… none.

Similar slaps on the wrist for posting government datasets and tables from UK and Scottish government data openly available to find, deleted for similar vague and unchallengeable reasons, just labelled as 'spreading misinformation.' A quite strange phenomenon of censorship and in itself a warning of encroaching our free speech and what happens if you query the mainstream narrative.

Chapter 10 - Jab or not to Jab that is the question?

At this juncture I want to briefly discuss some of the ethical principles I would like to think come as standard across medical and nursing colleagues.

Valid consent involves speaking to a patient about the full benefit of a treatment but also to be open and honest about the risk, to enable to allow concordance. A mutual agreement between patient and Doctor to enable a plan of action going forward.

For consent to be valid it must be fully informed. And even if I don't agree with a patient, as long as they have capacity, they are equally allowed to disregard your treatment despite your advice i.e., a valid refusal of consent.

Current medical practice is sadly lacking in the above in two areas:

1. The mandating of NHS staff to be vaccinated to remain in post. This is simply called 'coercion' and is not part of reaching the afore discussed consent to treatment. Coercion of an NHS member of staff to take a medical treatment against their will is simply, by anyone's definition, blackmail. From another perspective it is in breach of the Nuremberg

code and law: Public health act 1984 45 (C)

2. There is severe lack when it comes to some of the 'consenting' I have witnessed in clinical practice. Every human has a right to their own bodily autonomy. To give someone a treatment they refuse or to force physically or emotionally is considered, in law, to be assault and battery. To disagree with bodily autonomy and valid consent goes against the good medical practice that the GMC encourages us to follow.

Regarding point 1 above, as previously mentioned, age, comorbidity, role, and personal choice should all come into the equation. If the vaccine, as is proven the case by major swathes of data, does not reduce your ability to catch or spread the virus (i.e., reduce risk for others) and only confers (again open to debate) personal protection, then, to proceed, this is just unethical, unscientific, and implausible in so many ways.

The injustice and harm from this law and subsequent sacking faced is just simply not offset against the threat of the disease in its current state, particularly in view of safety question marks as well as efficacy of the vaccine itself, another large area of debate.

Many cases in literature of swift waning of vaccine immunity (particularly Omicron interestingly) is well documented. Conversely, in recent studies, natural immunity reigns supreme. And how about the detrimental effect this has on stress levels, mental health, finance and morale or our friends and colleagues facing dismissal, those self-same people who worked throughout facing this huge pandemic with such bravery?

Regarding point 2, simply asking a patient if they know they are here for their second covid jab, a brief check on contraindications and… stab… into the arm it goes… is NOT valid consent (I won't mention the lack of drawing back on the syringe to first check it is not in a blood vessel prior to administration used in all other IM injections).

Even on the conveyor belt of mass vaccination clinics this is just not good enough. People were presenting in good faith, for the greater good of humanity, accepting a treatment to the lay person which was unknowingly novel but 'needs must.' They are completely unaware that in that small amount of fluid injected, lies mRNA material that encodes a virus's genetic message in a variety of different transport media e.g. liposome coat and is novel and never before rolled out outside of a clinical trial.

Neither were they aware that it is being used under an emergency use authorisation (EUA) like a 'needs must' legal waiver, neither do they know that the major producers of the vaccine have, before they ever stepped foot in the lab, taken indemnity from prosecution for all current and future harm under the EUA running for many decades yet to come.

I could go on.

You may have noted, I haven't referred to the mRNA 'vaccine' as such, as prior to this pandemic mRNA as mentioned above had never been passed as being safe or effective for human use.

Play this back for a second… we are giving you a jab of fluid, a novel treatment, which will be absorbed into your human cell which, cutting out the detail of the biology, gets expressed by your human cell, using human cell machinery to transcribe that code into a protein which is how genes are expressed. A protein that is viral in origin and design. Not only that but a cytotoxic spike protein (as opposed to other areas of viral capsid) that will induce an immune response in a way never used before.

Previously vaccines were killed whole virus particles, live but attenuated, fragments of protein or their toxoid with all their retrospective relative safety data, in favour of an experiment.

I fully expect the same number of sleeves would not have been rolled up in knowledge of the above.

I have heard cries from my patients of 'I only took the vaccine to… go back to work, to go on holiday next year, so I can go back out and socialise at the weekend…' and tearjerkingly I heard of a report of a child who gasped with relief after his jab that he could now go and play with his friends.

No coercion here.

Nothing in life is without risk, even common medicines like paracetamol have associated risk. Given the Jab's roll out being a 'clinical trial' (this remains the case) why was it not made clear from the outset about how to report adverse events. It is something that has only very recently been made public, regarding yellow card and their use in reporting of adverse effects to patients in the national media many months after the roll out of the 'vaccine.'

This is just simply bad medical practice, however coincidental or frivolous it may turn out to be, surveillance of benefit is important, but surveillance of harm/risk should be in equal measure.

I have seen a multitude of blank faces and 'what's a yellow card' when asked if they have reported their relative's reaction to the government scheme. Even the ones who have, didn't persevere due to complexity or time constraints.

On that subject I truly believe, and again data is becoming apparent on the vastness of this underreporting of post vaccination adverse events/death with blind eyes being turned left right and centre. It seems to be, in every hospital and surgery, that these incidences are swiftly debunked by the medic or swept straight under the carpet.

Combine that with an equally difficult to quantify overreporting of deaths (I have personally witnessed testimonies of over fifty dubious COVID 1a) MCCD. The difference between yellow card deaths post vaccine around 2,000 and the death toll of circa 150,000 would be much closer by this bearing true and with each passing percentage point, narrowing of the gap, the risk starts to unacceptably erode the benefit!

I see a huge hesitancy locally and nationally to audit data regarding harm, ease in dismissing strange happenings to anything but the Jab. Data should be at the tip of the finger re how many cases of myocarditis in different age groups seen after Pfizer/AZ/Moderna, how many strokes or cardiac sequelae… click, click, result.

Who is collating this data on safety? Who is disseminating it? Or maybe the question should be, 'Who is hiding the data and why?'

Chapter 11 - Ethics

A final couple of miscellaneous points on the subject of ethics in clinical practice. Firstly, confidentiality is the cornerstone of medical practice, to speak outside the confines of the clinic room is a breaking of such confidentiality.

Patient's medical history is private! This seems to have been ignored particularly in the media, and among colleagues 'off with covid' and/or 'their vaccine status.' You only have to switch on BBC TV or Sky Sports News to hear about the latest COVID positive actor, politician, or sports player.

Do you ever remember switching on the TV and hearing about David Beckham's orthopaedic history, but its ok to reveal Cristiano Ronaldo's Covid status/shielding regime? Public figure or not, confidentiality is paramount in the profession but again, something vital has fallen by the wayside.

Finally, the principle of treat patients according to their best interests and in particularly counsel and treat based on this principle are so important.

I will end here with this. I am constantly hearing of many colleagues who are 'unwilling to have more jabs' and definitely won't be vaccinating their children… which is absolutely fine of course.

However, to have this view on a treatment you aren't willing to have yourself or recommend to your family but then proceed all afternoon to jab the children of thirty strangers has to be considered unethical at best.

The apathy is real.

Closing remarks

It is human nature to not want to admit you are wrong. It is human nature to not get too proactively involved in something that doesn't directly involve you. It is too easy to trust the government and scientists that stand before us.

But science changes, this pandemic changes. As scientists, medics, and the general public, we hope, as best we can, to keep updated. But to censor, close down, refuse debate or critically appraise the data around evolving harm/benefit profiling, is just not science, it's just not ethical or good medical practice.

If they mandate a vaccine (that they are already planning jab number 4) and so on, for NHS staff without respect for valid consent, without respect for the refusal of that consent then this is particularly important in reference to the arena of a clinical trial with no medium- or long-term prospective safety data.

Certainly, data suggests uncertain/patchy benefit in terms of reducing transmission i.e. protecting patients and certainly from recent data, is equally sketchy on personal risk reduction, that being the case then what is the point?

When do we say enough is enough?

When do we declare that the vaccine doesn't work or isn't worth the risk?

And when do we accept that mandate, aside from being unethical, is tantamount to blackmail in view of this?

The virus is now as good as endemic as per many media outlets report, let us have our bodily autonomy back, our freedom back, let us snap out of this 'mass formation' psychology of fear and regain our common sense.

There is a lot to say for the good old-fashioned herd immunity.

What happened to the banging the saucepans for your NHS heroes?

What happened with the value of the tests to show you haven't got the virus?

(Previous limitations discussed aside), PPE actually giving personal protection? And what about antibody tests?

As a member of NHS staff going to work to face patients, with positive antibodies proving prior immunity, daily negative tests, fit and well status and in PPE, then my conclusion is that this member of staff is no more a risk to their patient than their fully vaccinated counterpart following the exact same measures!

Primum non nocere.

This pandemic has been unprecedented for the careers of all those currently in clinical practice but if we aren't interested in this harm and preventing it, acknowledging it, discussing, and debating it, then we need to take a long look in our mirror before we next go on duty!

Chapter 12 - ITU shenanigans and end of life drugs

Following the viral spread of the above essay across the social media platforms, I was quickly thrust into a world that was uncomfortable for me. The public eye was on me as thus far in the pandemic concerns had been muted by the press and mainstream TV channels and dissent wasn't apparent or any debate to matters of safe and effective.

I embarked on a series of podcasts to spread the word to as many diverse audiences as possible, some large while others to small new podcasters. I wrote to MSM broadcasters to ask for live debate or interviews and never received a response to a single email. If your position is so strong it should be easy to defend especially against a 'crazy anti-vaxxer.' The key for me was to get this message out far and wide. I had started to have additional concerns about other aspects of the pandemic that were snowballing as time went on and wanted to vocalise them. Blanket DNAR (Do Not Attempt Resuscitation) policies rolled out, high dose and overzealous use and prescribing of midazolam and diamorphine. Revelations from Pfizer that the jab was never tested for ability to stop transmission. Ethical breaches such as jab and mask mandates and government overreach into personal lives of the populus.

As mentioned earlier my best friend Emma was the main catalyst to my own decision to take the jab. Her death after a seasonal flu type illness had been marked as covid and her death after twelve days on ITU was splashed over the press as covid-19. However, there was more to this story than met the eye as during her illness I was regularly updated on her care through my own enquiries and via her husband. Her stay in ITU came after she was unable to maintain her oxygen with non-invasive ventilation. I remember waking to a selfie image from Emma with a CPAP mask on looking ever so frail. I was advised to contact the hospital by her husband, and it became quickly apparent that she was to be ventilated. I called her just prior and could hear high flow oxygen and bleeping of machines and the nurse put me on the phone to her and prior to this was chillingly warned by the nurse to expect the worst and that this might be the last conversation with my best friend in the world! I couldn't muster any words through tears of what the nurse had unexpectedly explained to me just prior and all I could hear was Emma stating how scared she was as I tried to reassure her and told her that I loved her like my sister. Those were my last words to Emma.

Now, while I agree that 'Covid' or a respiratory illness led to her ventilation, her time on ITU was peppered with iatrogenic events that would highly likely have contributed to her death. She suffered two pneumothoraxes (punctured lung), likely in retrospect due to high ventilation pressures being used, she was de-intubated twice (practice at the time was placing patients prone and then on their back in cycles) the latter occasion was a prolonged down time and CPR was administered due to swelling of her larynx requiring a cut to be made into her neck to put the tube in through there (a tracheostomy). I am told that the doses of drugs used were high including midazolam and diamorphine which given she had a respiratory illness were well known to be synergistic in their ability to depress respiratory drive. Her death certificate I am told (as I never saw it) stated her medical cause of death was cellulitis, a bacterial infection that had been picked up late and no antibiotic treatment was able to be given in time. Covid-19 was further down the list.

This however didn't stop the press leaping on the tragic case of a young person succumbing to the disease after prolonged ventilation. Emma was a very unwell comorbid and immunosuppressed person, but this was not mentioned. Neither was any of the iatrogenic harm above discussed or mentioned. A scaremongering 'Mother 38, dies from covid' and 'Covid-hit mom, 38, told husband and 5 year old son she loved them in heartbreaking final call.' This reporting was a loud alarm bell for me at a time of grief of losing my friend. At the same time, I encountered numerous reported concerns of overzealous end of life medication prescribing and use, particularly in nursing homes. What was going on in my profession?

History repeated itself in a manner where I got involved in a similar case to Emma, where I spoke to a hypoxic lady from north Wales. She later became known to me as Aunty P, but essentially the story goes that she ran a viral upper respiratory prodrome of symptoms but developed a nasty cough and some shortness of breath. On my zoom call to her she appeared to be hypoxic (her lips were dusky) and her oxygen probe was showing 84-86% which isn't good. I advised her to call an ambulance due to hypoxia and that she could have covid/flu/pneumonia. The ambulance took 8-9 hours to arrive but once in she seemed to settle well and pick up with the help of some low dose oxygen delivered through her nose. That evening I received a text saying she was doing ok. The next day I received a distressed call from her partner that despite the improvement there was talk of ventilating 'Aunty P.'

I 'facetimed' her and saw her looking very well, talking in complete sentences with nasal oxygen in situ and saturations running at 98-99% with a patent airway. Not the findings of a moribund person in need of ventilation. In fact, she joked that she was bored and had asked for the TV in her side room to be fixed! I called the consultant on for ITU and enquired why she was being ventilated. She presented to me an ever-improving set of blood gases aka she was recovering with low oxygen supplementation, her chest x-ray showed mild infection changes, and her infection markers were only slightly raised. Non-invasive ventilation had not even been considered (the step before ventilation with tight fitting high flow oxygen masks). The doctor was adamant that they needed to follow what she referred to as and for the first time I had heard 'the covid protocol.' This protocol remains a mystery to me other than the consultant telling me part of it was 'pre-emptive' ventilation as they 'knew' given her positive covid test that a deterioration was 'inevitable.' From ITU anonymous whistleblowers, I have been told that other parts of the protocol are high doses of synergistic respiratory depressants midazolam and diamorphine enough to tranquilise an elephant, and immune-nuking doses of dexamethasone in addition to concerning blanket DNAR completion policy across the board.

This was an alarming development for me, and I wondered how many covid ITU deaths were iatrogenic from pre-emptive intubations, harmful drug cocktails in addition to ignoring the contribution of iatrogenic harm in Emma's case sensationalising the fact that COVID-19 was the sole cause of death in her case was simply not accurate. On what scale were these two case studies happening and to what extent was this driving mortality statistics and fear?

One further concern to document at this juncture was the practice of completing treatment escalation plans. At a time when people were limited to small 'bubbles' of their own family to be in contact with and lockdowns merging into one another, GPs were being tasked to visit the elderly and infirm in residential care and independent homes and make a judgement call. This was usually done without the patients or the relative's input and decisions as to what level of escalation of treatment would be suitable if they were to contract covid or become ill from other things. The main result was to deem out of hospital treatment the ceiling for the elderly and upon completion of this decision documentation, a blanket DNAR (do not attempt resuscitation) form was completed in addition to, in many cases, a pre-emptive 'just in case' end of life cocktail of drugs being prescribed in case end of life care was required on deterioration. This was all so hypothetical and surely each case needed to be discussed independently on merit rather than such a blanket policy. There have been published concerns from other whistleblowers that as a consequence of drugs being 'ready to go' that overzealous and premature usage led to some deaths that had also additionally been documented as covid-19 based upon the previously discussed flawed test results. I have certainly witnessed this zeal for syringe driver commencement particularly in residential setting and out of hours periods.

One particular case that lives with me was over the Christmas period. I was called on Christmas eve for a late visit to a previously well gentleman who had become delirious, confused and had some incoordination in his motor function. I examined him and he had a fast heart rate and high temperature but was looking well in himself. On systems review and after full examination, it was clear that he had developed a perianal abscess and was showing signs of this infection. It was discharging well so I dressed it and prescribed antibiotics. On my return to work after boxing day I set up a telephone call review and things had improved, and his daughter stated he was back to normal. Later in the week he took a turn for the worst and a visit request came for a review. I was working at a satellite surgery quite a way from him, so a colleague decided to assess and felt that this deterioration was due to the perianal abscess reflaring and extended the course of antibiotics. I had a few days annual leave after this shift and upon my return this patient had been notified as a death!

I was utterly alarmed by this and read the notes. In the interim day's he had become increasingly delirious, and a senior doctor had been out to assess completing a DNAR and discussed keeping him out of hospital and prescribed pre-emptive end of life drugs and continued with add on antibiotics. The next day he passed away. The death certificate was marked as old age! The family even thanked the surgery for their kind care of their dad in the last few days of life! A previously independent healthy man without any comorbidity, seventy-two years old with a surgical problem and sepsis related to this and needing admission, was dead. I wondered how many similar cases were happening up and down the land.

Chapter 13 - Anecdotes of covidian lunacy

I wanted to share a few stories from my clinical practice of things that happened during my time throughout the scamdemic that didn't naturally slot into other parts of my diary that caused me great concern regarding the mentality of the profession and its lost ethical compass.

First a story of one of the last surgeries I had worked at before the affirmative blacklisting. I walked into this surgery recognised by the senior partner and several members of staff for my stand against the narrative. This was one of the only times I had ever felt accepted for my dissident views. Why are you still not working there? I hear you cry!

My initial time there was good, and I got back into the swing of general practice after some enforced time away and soon my confidence was growing. As my time there lengthened, I started to have more coffee room chats and soon found out that one of the reception staff had died of a stroke very shortly after the jab. I heard about a member of staff that was off with what sounded like an ever-growing occurrence of turbo cancer that I was seeing in the young, jabbed patients. Occasionally, my door would tap with a sheepish doctor entering and locking the door behind them, to enquire with a whisper, being sure not to be seen with the heretic, for advice as to how to help a couple of his patients that were likely jab injured. This was all progress.

Sadly, things turned sour after a positive start. I was made aware that three young nurses have all suffered either myocarditis or pericarditis after one or the other of their jabs and had collectively refused to take any more injections. During a morning meeting the practice manager out of nowhere and with the context that this was what I thought was an 'awake' practice proudly announced the Spring 2024 boosting booster booster campaign! I felt sick! To add to the sickness the three refusenik nurses were in charge of giving the jabs to the 'at risk' cohorts. I really couldn't reconcile the concept of a person refusing a medical treatment for themselves due to adverse events being willing to administer the self-same medication to a patient. Towards the end of my time there, two clinical encounters stick in my mind.

A middle-aged man wanted his flu jab but on arriving to his appointment the nurse said she could only give the flu jab at the same time as the covid booster. He was booked in with me to discuss this after refusing the offer. Of course, I agreed with his stand that not only was there no clinical trial safety data regarding the co-administration of a flu and covid jab, but this was unethical coercion. I backed the patient and asked him to research the flu jab independently and to return if he was happy to proceed.

Two days later I saw a young child, six years old. He had come in with his dad. The presenting complaint was nonspecific and based around a month of general malaise, lack of energy, poor appetite, and a small weight loss. After a full systemic review and examination, I looked at the screen for medical history stumped by the cause of this decline in a previously healthy and energetic youngster. I asked did anything change before he became unwell. Dad thought for a moment and recalled that it all started a day or so after his flu spray. I looked at the notes and confirmed the date of administration. More alarmingly I saw an entry in the notes documenting a lot number and a same day administration of a covid 19 jab. I calmly (despite my anger) enquired about why his dad had chosen to give a five-year-old a covid-19 jab only for the dad to appear equally perplexed. As it turned out his dad was a covid refusenik and I literally watched his face glow with anger. He stood up and exclaimed, "You mean to tell me that my son has been given a covid-19 jab without my consent?" to which I turned the screen to him to see for himself. He clutched his son's hand and made a beeline to the reception desk asking to speak to the practice manager. Sadly, this was the last straw for me. I am not aware of the outcome of this incident, as I had resigned the very next day on ethical grounds, despite this being an awake surgery and a very financially lucrative arrangement.

Another tale of patient safety and ethical practice leaving

the building. I worked back in the Midlands for a short working holiday at one of my old practices to cover some annual leave. Since one of the partners had suffered from myocarditis after his first jab the surgery had slowly awakened to the possibility of jab injury, and I was always vocal about my position. They kept having me back there despite this but sadly continued to jab jab jab! The practice manager sheepishly came to see me on the lunch break of my first day and I asked her why they were still jabbing patients. She pinned all responsibility on the other senior partner, a vaxaholic, maskaholic who jabbed anything with a pulse. She came to me concerned about one specific situation. The practice was delivered a batch of the latest boosting booster, and she was alarmed to see that the vials had printed on them an expiry date that had elapsed even before the date of the delivery! She told me that she contacted the supplier and was told to use a black pen and scrub out the expiry date with a thick black line and to use the batch until completion and this had been agreed locally that expiry dates did not need to be adhered to! Insane.

Finally, a story about the local coroner. A local family asked me to attend a postmortem of a wife and mother who had recently passed away. She was fit and healthy and if I correctly recall seventy-two years old. After receiving her first covid shot her health took a turn for the worst. She went from healthy and independent to almost wheelchair bound within a fortnight due to a constellation of rapidly progressing symptoms after feeling 'not right' immediately after the jab. Her uninvestigated symptoms included abdominal distension, breathlessness, intermittent confusion and rapid weight loss and general weakness/lack of appetite. When she was called for her second jab the family explained the decline in her health since the first. The surgery used this as further justification to have the next one because she is more at risk and vulnerable to covid.

Her family reluctantly agreed with this logic and wheeled her, chin on chest, to the surgery for her next jab. Days later she passed away.

I attended the day of the postmortem and met the coroner who stated, "I thought I was the bad boy of medicine until I heard about you," and shook my hand warmly. This was an encouraging start as he went on to have a brief discussion about my position and was in complete agreement with my stand. We proceeded to the postmortem to discover a plethora of pathology on gross inspection including fluid in the abdomen, masses on her gynaecological organs and in her chest. Part of a postmortem involves taking sections of major organ tissues for examining under the microscope called histology. While the doctor was preparing the various samples, I enquired from the coroner whether he thought this was a jab related death and his response was guarded. He stated that jab deaths, "Couldn't be proven," which I struggled with given this was a post-mortem to determine just that.

I had recently become aware of a pathologist Dr Arne Burkhardt who had been doing postmortem studies and published work on proving spike protein depositing in the testes, hearts, and brains of young, jabbed people who had died suddenly. This was a reproducible observation across the series of investigation and consistent. I asked the doctor if he knew the stain that he had used and whether the NHS was using this, how else is a coroner to discern jab related cause of death. Again, a case of if you don't look you don't find. He stared blankly at me almost in disgust and proclaimed we don't test for that in the UK.

We continued the post-mortem to completion and later heard from the family that the inquest concluded a death from natural causes despite the temporal decline in health after jab one and death shortly after the second. This was an outrageous verdict and exactly the reason that what I consider the biggest medical negligent event in history and subsequent cover up continues to this day.

The same coroner filed a death of a young man with multiple blood clots dying in his fathers arms (featured in my podcast, Breaking the Silence), having been fit and well prior to a rapid health decline after the jab under 'natural causes' with the grieving dad unable to get any explanation or answers, his emails perpetually ignored. Utterly tragic gaslighting and I wondered how rife this frivolous and inaccurate documentation of deaths was happening in lockstep with coroners.

Chapter 14 - Friends old and new

The medical blacklisting of me has been discussed elsewhere in this diary. I wanted to briefly discuss what I know for sure is a universal reaction of past friends, colleagues, and family, towards a dissenting covid voice. This has been unprecedented and visceral.

Personally, I have experienced a torrent of abuse directly and indirectly online and in real life by a small number of ex-colleagues. The rest of my seventeen years' worth of working colleague acquaintances have literally shunned me. Many family members and friends outside of medicine have cut ties. The cancel culture towards any notorious 'anti-vaxxer' was a zero-tolerance to their 'selfish views' and as if they have a bad case of leprosy deserving of isolation. A black sheep among friends and family!

I know many people reading this will empathise with my experience of losing friends and family ties, sometimes even those who were very close connections before the pandemic. What ensued was a mass formation psychosis, fuelled by the media and many celebrities who took to their platforms to demonise the dissident voices and jab refuseniks. I have heard many reports of others being shunned from family events such as weddings and funerals as well as not being invited to Christmas dinner due to their vaccine refusal.

Despite this, and questioning 'were they really friends to start with?' if you can be shunned for a difference in medical and scientific opinion. However, in these strange times, I am sure we can all attest to finding our 'tribe' in among the madness. Almost two versions of reality ensue as we seemingly coexist within a parallel universe of vaxaholic covid cultists and safe and effective brigade fixed and firm in their beliefs identified usually by their mask wearing antics. It is perhaps important to state that a fixed and unshakeable belief despite firm evidence to the contrary is medically diagnosed as a delusion. I can safely say that despite my preference of being a lone wolf, I have met some inspirational, brave, humanitarian heroes both within and outside of the medical profession. To name them here would be a long list. Yet safe to say the list simply isn't long enough in the expectation that people by now should be able to see the evident problems of the jab, pandemic policy, and its collateral damage. Though very few have spoken out due to fear, home comforts, financial motivation, corruption, or combination of any of those.

Many 'awake' groups have been born out of the scamdemic era, and I consider these people, online and offline, as my sanity and support. Real friends and adopted family as we all seek sanity in a world suffering mass formation psychosis. I just want to thank every single person who has supported me from those who penned complimentary social media posts with words of thanks, social media contacts, and telegram groups in addition to what I now regard as my new family. You have been the light through the darkest of times. You know who you are but never underestimate the power of what you did and continue to do.

How will things change I hear you cry? Well, my gut instinct is that there is a tragically silent realisation among the profession and the multiple-jabbed as to the grave error of their decision to take, promote and administer what is clearly something that has sinister implications for humankind. Whether it is their own deterioration in health, the rapid onset of cancer or disability within friends and family circles, or the frequency of funeral attendance following the death among those in our communities, slowly the stark realisation of the common denominator among the dead and the injured, the elephant in the room, that which must not be spoken, is a correlation slowly being made.

I know this for certain as a growing number of colleagues are frequently contacting me via direct messages commending me on my actions, confirming their agreement with my position and that there is a major problem becoming patently clear. Likewise, one of my common health coaching appointments are the awakening multitude of patients coming to this realisation for themselves. Having been gaslit and shunned for daring to make the link, they come to me for a listening ear and advice. These poor victims are not best pleased with the profession, and this is where duty of candour (covered later) is now imperative.

I recently had a reconnection with a medical school friend who had initially blocked me as I had advised them of the risks and lack of safety data in pregnancy and to warn his partner to not take the jab. What upset me the most was that we were quite close, and he used to always joke about what a 'geek' I was with my studious nature and time spent in the medical school library. The shunning was hurtful. I have not heard from this chap for three years until he reached out and said that my situation with the GMC and tribunal had come to his attention. He called me, offered me support and commended me on my bravery. I asked him in his time as a now colorectal surgeon whether he and his colleagues were seeing increased cancer, whether these were in younger cohorts than previously and if he was seeing any issues with mesenteric circulation clots. He confirmed my fears that younger more aggressive cancers, unusual clotting, and intraoperative bleeding, were rife, and people were now starting to make the link to the very real connection to covid jab status. I enquired as to what action he and his colleagues had taken to alert the profession to the observations, and I was met with silence and a look of sheepishness. I asked what his plans were, and he confirmed, due to his observations, that he had left the profession and is working as an entrepreneur in the surgical equipment industry, choosing to walk away from the controversy rather than raising the alarm.

Even churches were not exempt from the insanity. At a collective time of spiritual need in the face of ramped up fear, churches slammed their doors in the faces of their flock, reverting to online worship during lockdowns. Many people left the church due to this or were being unable to enter the new digital church age. When churches eventually opened, I remember walking in to find yellow and black hazard tape stretched across the pews and signs reminding of two-meter rules and even the urinals were not safe with every other station criss-crossed with tape. While 'socially distancing' throughout sermons people were encouraged to leave swiftly, not congregate, and the post worship cuppa was contraband. Wearing masks and visors was the norm as more and more obscure rules and guidance was proffered by the government only for the church to unquestionably obey. One strange moment I recall was the pastor saying that we had to always wear masks unless we were singing at which point they could be removed. If ever a useless face muzzle was to be of any fractional value would be the aerosol generating action of singing praise! At this time, my fellowship was starting to fall ill, and I watched as illness and disability ran amok among the congregation. I tried so very hard to raise the alarm but was continued to be reminded that we were there to worship and that my views fell outside the remit of the service or informal discussions. The place of worship and of sanctuary became the place of government compliance.

In Roman's 13:1 there is a scripture which many churches often quoted to justify the odd decision making and blind obedience. 'Let every person be subject to the governing authorities, for there is no authority except from God, and those who exist are established by God.' In other words, submit to authority given the divine origin of said authority and consequences of resistance. However, ignorance to the fact of some of the more spiritual aspects of the scamdemic for example harming unborn babies, giving pregnant women gene-based injections, and ignoring a groundswell of jab harms were being ignored which I always felt was an unusual position for the church to take.

There is a higher authority for when government overreach and criminal negligence is being committed. God's authority ABOVE the aberrantly behaving 'authority' especially when much of what has happened in the last four years or more is prophesised in scripture being fulfilled in events that were coming into reality. Where was the recognition of this and the spiritual battle ensuing. Sadly, it was in the same place as the medical professions outcry. Nowhere to be seen but for the brave few outlier churches who took a stand and kept their doors open throughout.

Chapter 15 - Cathartic therapy

As you will probably understand by now, my stand against the narrative has led to a lot of stress, trauma, anxiety and insomnia for me and my family. My medical training and the regulatory guidelines I follow strongly encourage safety in whistleblowing, recently rebranded, ironically, as 'freedom to speak up guardianship.' My stand was very basic and focussed on asking questions about the science being used to justify many of the pandemic measures, the ethics around coercion and lack of informed consent, my concerns about the requirement for poor quality masks and collateral damage of these as well as lockdown policy on the economy and on mental health. As the age of cohorts dropped to include young and healthy adults and subsequently children and pregnant women, the line was crossed. Surely my colleagues could at least agree on the lack of clinical indication for these cohorts, given the miniscule infection fatality rates in addition to lack of any pre roll out safety data, followed by very public claims of safety and efficacy. Pregnant women and children were straight off the bat exclusion criteria. Surely this was a no brainer?

My final crime against the profession was daring to equate correlation and causation to vaccine injuries and daring to help them. My profession was more than happy to label any sequelae of post vaccine syndrome presentations without any evidence to all encompassing 'long covid' to explain all. This was something applied without question, yet the same cohort was simply unable to attach attribution to someone having a blood clot the day AFTER the jab or a stroke to a potential jab reaction. It was simply never tolerated to make this link. I cannot remember a single conversation or debate with a colleague over the last three to four years that has provided any solid evidence to rebut any of my concerns. My appraiser at the time commented on my annual appraisal (in the integrity/probity section of the assessment ironically), that while my efforts at whistleblowing were to be commended, that I needed to be reminded of my commitment to 'follow NHS consensus opinion' and that I would have been better served to have (and I quote), 'Kept my head down and shrugged my shoulders.' Chilling and disturbing given the medical seniority of the author.

As an example of medical gaslighting, I was working at a practice where I saw a patient brought in at the end of the morning surgery, by her mother, in a wheelchair. She was PEG fed (a tube placed externally into her gut) and had dysphasia (slurred speech) and a weak left side. She worked in healthcare so would have been coerced into taking the jabs to keep her job by taking a jab that had no data to show that it stopped transmission. She took AZ as her first jab and developed a small deep vein thrombosis shortly after.

This was managed with a short course of blood thinners and the vaccine was exonerated from causality. This was demonstrated by the fact that despite a likely provoked DVT from the AZ jab, she was given ANOTHER AZ jab which led to another much larger blood clot to her other leg! At this point there had started to be some discussion about a possible correlation between clots and the AZ eventually culminating in market withdrawal. She was then asked to get her booster, this time the vaccine centre mixing things up with a 30th birthday present of Moderna two days after her big day. Within twenty-four hours she had a mini stroke known as a TIA. Again, no attribution despite the temporal correlation to the jab. One full week after the jab she went on to have a large right middle cerebral artery clot aka a stroke leading to her losing motor function on her left side (the brain is a mirror image of which side correlates brain to muscle function). I saw her some months after this had happened for something quite unrelated.

I asked her what the doctors felt caused her stroke and I remember this moment very well. She dead pan exclaimed that the events were down to clots caused by 'asymptomatic long covid.' I nearly fell out of my chair as the anger raged in my mind. She had been genuinely gaslit to believe this despite confirming that she had never had any symptoms of covid, nor tested positive in her history yet had been convincingly sold the theory that at some time in the past she must have suffered a bout of the 'deadly' virus without ever knowing it and no test to confirm. Outrageously, this claim was documented in her neurology outpatient's letters and discharge summary and was coded in her GP notes! This story is one among many of much evidence of medical gaslighting and anti-science that I have discussed elsewhere in my diaries.

I rang the senior GP on call when I had a few spare minutes before phones had closed down for the day and explained my dismay. We had a circular conversation whereby he supported the claims of asymptomatic long covid despite lack of evidence yet bizarrely used the argument that 'correlation doesn't mean causation' that we have all heard of despite the temporal 'coincidence' he claimed and asked me how I can prove that it was the jab. Welcome to my new world of medical ignorance and the first signs of prion disease in my colleagues academic reasoning and critical thinking.

Following our altercation, I was not invited back to work at the surgery.

Between roles I suffered from a prolonged bout of insomnia and decided that it would be a productive therapy for me to pen a poem that summed up my thoughts of events and subsequently led to me reaching out to fellow 'dissenters' eventually leading to publishing a collation of poems written by myself and other refuseniks. I entitled the book 'A doctor's despair' as at this point, I was despairing at my profession's silence, gaslighting complicity, arrogance, and cowardice! I wrote this poem at 3am and had been planning it in my head the hours prior while tossing and turning in bed.

A Doctors Despair

The dark days of these last few years,
A new war, no need to fight,
Totalitarian legacy,
A battle of dark versus light.

A new disease sweeps through the world,
The media swelled the public fears,
That broke a generations psyche,
With a multitude of tears.

Day after day the psyops grew,
Of death, despair, infection,
Apocalyptic imagery,
A solution… gene injection.

Locked in, shut down, alone in homes,
The elderly in isolation,
Images of palms on windows spread,
Accepted by the nation.

The fear of God into our kids,
Governments enforced the rules,
The tears of all our children's eyes
Masked up, sent home from school.

At warp speed, solutions found,
Lockdowns, masks, distance, big pharm,
The speed of science, before we paused,
Injections into arms.

It all seemed rushed, corners cut,
Ethics and safety swept aside,
The solution in a Pfizer vial,
Turned into genocide.

The harms became apparent,
When the roll out gathered pace,
Injured, hurt, perpetually ill,
The despair on my patients face!

Reflecting now I can't explain.
The overreach and violence,
Displayed by those charged to keep us safe,
But met by colleague's silence.

Where is the anger, where is the rage?

The public lost their fight,
Our children's future met with apathy,
Darkness seeming to defeat the light.

But when all hope seems to have been lost,
A small remnant awakens,
In isolation a few stand out,
The fire in their souls are shaken.

Against all odds, Against the flock,
In isolation first, against mounting derision,
An army of brave soldiers build,
To fight totalitarian division

There is a hope, there is a dream,
That we beat this reigning fraud,
Stand up, fight on, small victories count,
A critical mass aboard.

It's time to stand, fight back be free,
Resist the great reset,
A new world order nobody wants,
Look back with no regret.

Chapter 16 - Maskaholics

Masks were a major concern for me during the scamdemic. I recall with disdain trips to the supermarket that stuck with me as I wondered at the majority of the people in the store who were trudging around, faces covered with masks in a zombified state of expressionless marching down the aisles. It was very bizarre and apocalyptic seeing their waxy staring expressions as they robotically filled their trolleys. I commonly witness exchanges of masked pensioners arguing with mask dissidents for their selfish refusal to wear masks in addition to 'mask around chin' syndrome as well as mask wearers sat in the cafes frequently pulling their mask away to stuff a large, iced bun into their mouths washed down with a can of coke. Health conscious all the way.

Early on in the introduction of face masks some of the boxes that dispensed them in my clinic room clearly stated they are not for protection against coronavirus, yet supermarkets, social venues, and workplaces intermittently mandated/strongly encouraged wearing something that simply didn't work. Additionally, to lacking efficacy they were just not safe. Studies were being published about links to microplastic exposure, increasing carbon dioxide and reduced oxygen levels even after short term wear (hence the spaced out stares of the maskaholics), and even a disturbing paper from the journal Nature which swabbed the inside of a surgical mask at the end of prolonged wear to find a petri dish of bacteria and fungi that people were expected to wear to 'protect' themselves and others. Seeing children wearing them in school, stood distances apart, made me despair as to the mental health impact this would all have in the future. Especially paired with the media propaganda that they were living through a deadly apocalyptic virus pandemic straight from Netflix!

I recall meeting up with a friend for some lunch and as I entered the restaurant maskless one of the bar staff stopped me in my tracks and asked me to wear a facemask. I didn't have one on me and was asked to leave. I went back to my car and found a very old, sand and mud covered crusty old mask in the footwell of the car, and I decided to toy with the staff. I re-entered the restaurant with this filthy rag on my face and was allowed to pass through the bar into the dining area to meet my friend who was already seated. As I entered the room the waitress said that as I was dining, I could remove my mask while seated. The protective effect of a seat and steak meal from a deadly virus! Things like this made a mockery of common sense something I later rebranded 'covid logic.'

I worked locally at a service that had recently reintroduced masks as a mandate in certain parts of the building front of house. This in itself made no sense aside from them not working as previously discussed. I started my shift in a packed department and in two hours was due to be the only doctor covering the shift. I hadn't worn masks for a very long time and in my three weeks prior to this fateful day had clearly been working mask free on shift. I saw my first patient and while writing up the discharge notes was tapped on the shoulder by the nurse in charge. She asked me for a word and after completing the paperwork met up in a side room. I was politely, at first, requested to wear a face mask and reminded that masks were mandated from that day in that area. I calmly explained that it was a hot day, I had asthma and didn't want anything over my face in addition to headaches that I used to have at the start of the scamdemic when I was naïve enough to fall for it. Despite this the nurse persisted to require me to wear one else I would be asked to leave the shift for defying trust policy. At this point it was time for a more evidence-based angle. I always keep files on my phone for moments like this with 'at my fingertips' published journals to supply some facts. I presented her with three papers, demonstrating blood oxygen and carbon dioxide changes after short term use and the 'petri dish on the face' article mentioned above. Faced with evidence against her position she called me belligerent and advised that I would need to leave the shift for refusing the muzzle.

I enquired what would happen to the near twenty people waiting for assessment given I was the only doctor on later in the shift and she responded by saying that the unit would have to close rather than me remain to ensure the safe management of several elderly patients and a couple of infants waiting. I left the shift as had been threatened with security removing me from the building if I didn't leave voluntarily. Covid logic!

I was suspended for two of my three days a week from working due to this but in my other role I was allowed to continue at a satellite clinic in the community as mandates were only in certain parts of certain buildings. Covid logic. I had been off for a few days and started the morning with a virtual ward round on zoom. On the call were three other people, a manager and two doctors working from home. One of the doctors had a mask on in their home office with no one around and noting this I took the opportunity to explain what had happened the day prior and the lunacy of wearing a mask at home alone! I don't recall being rude or aggressive, the ward round continued, and the rest of the day was uneventful. The next day would have been my next shift but for suspension, but I noticed an unknown number ringing on my phone. It was a very senior medical doctor calling to 'see how I was' as there had been concerns expressed about my mental health following a 'rant' about masks during a ward round! My jaw dropped. He wanted to check if I was coping and proceeded to condescendingly assess my mental health. He concluded I was fine and made no comment on the mask situation. I left this role shortly after. Covid logic had descended removing any common sense from being applied by maskaholic medics!

I wrote the following poem to try and express my concern about the impact of these useless rags shortly after again as a form of cathartic therapy; this was also included in my anthology:

The Scars of a Mask

The power of a smile,
The anguish of a frown,
To tell someone you're feeling great,
To signal when you're down.

We smile when we are joyful,
We frown when things get tough,
Our faces show the world our thoughts,
Our grimace when we feel rough.

The barrier to this — the facial mask,
The debate of whether it works,
To stop the spread of covid? No.
But yes, to displays of joy and fear and mirth.

To see our children suffering,
In an emotionless world of fears,
To see our children reaching out,

A masked child's scars bring tears.

We rely on the power of our faces,
To tell the world we're fine,
We rely on the power of expression,
To show all the feelings of mine.

It's time to lose the facial masks,
For the good of humankind,
So, we connect again as humans do,
And display our souls and mind.

Let us see in years to come,
When we hope we've won this race,
The needs and dreams of all the world,
Freely etched upon our face.

Chapter 17 - The social media fallout

Thus far, you have heard anecdotes of life as a doctor and my transition from someone initially fearful of a deadly virus ripping through the world who stupidly took the jab and even wore the odd mask at the beginning to a more critical thinking dissenting voice who started to ask questions and research an evidence base.

What I didn't realise, naively at the time, was that dissent was not to be tolerated. The scale of this was unknown to me as I expressed genuinely held heartfelt concern with what I felt was a solid evidence base focussed entirely on patient safety. I had continued to work regularly as a locum doctor and in the background continued to speak on podcasts, conferences and speaking events, as well as growing my social media presence.

There seemed to be a groundswell of 'awakening' for want of a better phrase in the initial parts of 2023, as governments continued to plug booster after booster and increasing incidence of vaccine related adverse events. People were slowly and collectively reaching their 'eureka' moments in and out of my clinics.

My first pangs that something wasn't right was censorship online. I had LinkedIn, Facebook, Twitter (now X) and had recently set up a You Tube video channel to post my interviews and planned to do some vaccine injury case interviews. Facebook was the first platform to start applying sanctions to my account when I posted peer reviewed journal articles or government data sets against the narrative. My posts were shadow banned, tagged with warnings of misinformation or simply removed. This was unprecedented.

At the same time, I had started to grow in follower numbers on Twitter with many posts going viral and being shared. At this time, I was aware of shadow-banning, but my first realisation that something sinister was afoot was when I encountered what can only be described as coordinated trolling. A group of mainly anonymous accounts reporting my content leading to short term sanctions and firing ad hominem attacks at myself and any other dissenting online voice. Always abusive, hunting in packs and with very little, if any, data, or studies to bring to the debate, playground name calling was all they had in their locker other than what seemed to be a special influence over Twitter in the pre-Elon Musk era of the platform. Accounts were shut down sequentially and what seemed abundantly clear is mass coordinated reporting by this group played a part. My original account with a large following was shut down without explanation and following brief stints as 'DrJabjabjab' and 'Drboostboostboost' reincarnations reaching a similar ban I gave up on Twitter until after a long period and Elon-Musk taking over the platform promising uncensored free speech and out of nowhere my original high follower account was reinstated.

As my followership grew in the new era the trolling took on a more progressive and sinister tone as they stepped up their game of abuse and censorship. Anyone who showed me support was piled on in a similar fashion and the accounts perpetrating this seemed to be untouchable in their behaviour, while zero tolerance to covid dissent ran amok. To this day this troll network with obvious links to the industry they serve, big pharma, continues with a large number of accounts. Recently I have had to report them to the police after initially hoping that they would just leave me alone.

The reason for this is that there were a number of fixated and obsessed trolls obviously given the task of bringing my career and reputation down in a coordinated and malicious manner. In the last year I have experienced coordinated police reports (all thrown out), cyber and real-life stalking, doxing, hacking, mass weaponised GMC reporting, contacting my employers and friends as well as daily bullying, abuse, and harassment. It has seemed unrelenting; it is clearly funded due to the sheer numbers involved, effort, and time invested/wasted by these obsessive people. That is all the airspace they will get in this diary.

I had started to get some traction on LinkedIn a more professional platform where more civilised debate was occurring and my papers, journal articles and data sharing was seemingly uncensored. I did an interview on a podcast where I was asked to dig into the Pfizer trial data on the incidental pregnancies that had happened (pregnancy was an exclusion criterion but some who took part didn't realise they were in the family way). I went to the trouble of awkwardly holding up documents fully referenced in the video, running my finger across the pages and analysing their published data showing damning outcomes on foetal mortality as well as adverse event frequency and severity being unacceptably high. I was quite proud of this interview as 'misinformation' could not be levied at this detailed but concise expose. After all the documents were Pfizer and Moderna's own SEC submissions. These two documents also detailed how these jabs were distinct from being vaccines and were 'immunomodulatory gene therapies' and were unlikely to enter the market due to poor performance in prior clinical trials and other regulatory issues. These documents were filed in a matter of months before they started jabbing humans as part of the roll out. A technology never ever before used in humans outside of (poor outcome) clinical trials using a novel gene-based process whereby the human body was being given an instruction manual to make a foreign protein mooted to be part of the pathological spectrum of the covid virus… the spike protein. Absurd. Particularly as Pfizer had initially attempted to hide their trial data for seventy-five years only for a court ruling to be made to refuse permission to Pfizer who had to release the information including damning biodistribution data and awful trial data as well; bear in mind that they secured blanket legal indemnity from any vaccine fall out.

The reason for mentioning this interview is that I proudly circulated it across my social media and at this point started a number of other platforms including Instagram, GETTR and Truth Social. Within minutes of this interview being posted on Facebook, it was removed. Instagram, removed. YouTube, removed. Twitter, swarmed by trolls and eventually 'fact checked,' but my main dismay at sharing this interview was a permanent suspension from the LinkedIn platform which not only seemed harsh but was without any explanation. After months of trying to find out why I eventually got an email back stating I had broken their community policy on 'misinformation' and that the ban stood; no appeal and would be lifelong. LinkedIn, is that still the case? Are you still unwilling to allow facts to be aired?

What on earth was going on for this lock step of social media censorship? Simple, big pharma was what happened. An obvious investment against misinformation, otherwise known as truth, which may just bring the spotlight on the 'safe and effective' narrative and adversely affect profits. In a sane world the facts would be so damning that those in medicine, pharmaceuticals, politics, and the mainstream media, who pushed mRNA technology, would be in jail for the rest of their lives.

Instagram decided to put a banner on my account which stands to this day asking potential followers if they are sure they want to follow me as I apparently consistently breach their community guidelines. Despite this my follower numbers grew so I guess people know for themselves and I consider this banner good advertising. Any publicity is good publicity and all that.

My final flourish into social media came when I had the idea to interview rather than be the interviewed, focussing on giving some exposure and a platform to vaccine injury victims and their case histories and experiences. I recorded my first on a healthy man who had crescendo mini strokes followed by a large stroke disabling him. His wife spoke for him as he says in the background, he found it difficult to speak. I uploaded this onto YouTube as I felt that channel would give a voice, finally, to the jab injured and bereaved who were being ignored by MSM. Just one hour after this upload had completed, I received an email stating that my account was going to be locked down for breaching community guidelines for a week for posting 'misinformation.' I was deliberate in my interviewing to not be leading and allow my guest to speak about their lived experience and their own impression as to jab involvement. How could this be deemed misinformation? Who was defining what is deemed misinformation?

Not to be put off my plan by this latest setback I continued in earnest to record another six videos. I eagerly awaited the seven day ban to elapse and within minutes of its reincarnation rapidly uploaded my latest six offerings to the platform. Initial viewership was being registered, and I went to bed escaping the rapid ban dished out for my last crime. However, this victory was not going to last long. I woke up to a message from a friend saying the video that I posted cut out halfway through her watching it the previous evening. I checked on YouTube and the link to my channel had stopped functioning hence my videos were inaccessible. I checked my email and had been advised that I had received a 'second strike' and was banned from uploading further misinformation that breached community guidelines despite the latter videos having very similar content of lived experiences of jab injury being discussed.

In the following months as the truth was beginning to be revealed to the world, censorship eased. Despite promises from Meta that freedom of speech and censorship will be stopped, this has not come to fruition. However, X continues to be a beacon of freedom of speech and expression, the downside of which is the gang of fixated trolls tweeting negatively each day from their dark, damp cellars about me and others.

I continue to publish these stories despite the weak attempts to stop me and have proudly started a Substack/podcast where I continue to interview vaccine injury victims but with a new and unique content of NHS and scientist whistleblowers adding their voices to the debate, 'Breaking the silence: the conversations.' I will never stop; I will not be censored; the truth will set us free. Given my social media following on X has gone past 300K and across all platforms over 370k, what this little old crazy 'anti-vaxxer' GP has to say has the support of the people and that support for truth is growing daily.

Chapter 18 - Jab injury and the gaslighting brigade

In my career, I am not aware of encountering many cases of vaccine injury and was completely unaware of this as a concept. I was a product of the medical school induced 'safe and effective' mantra and understood the subtle warnings to never question or prepare to be 'Wakefielded.' Hence, by conditioning, screening patients for such presentations was never on my radar. I have advocated this message to my patients, never questioning, effectively leaving this aspect of medical practice to vaccine nurses in primary care, who, if they were doing their job properly would constantly be reviewing the evidence for safety of any and all vaccines they intend to administer.

Education en masse has been replaced with indoctrination and medical teaching is no different; while vested interests fund training, putting the patient first has long been replaced with putting the profit first while absolving themselves of any crime. The dilemma is that without constant questioning 'Science' becomes, 'The science,' 'Truth' becomes, 'Our truth.' George Orwell would recognise the direction of travel of this world and for the sake of our kids and our grandkids we need to stop 1984 becoming fact.

The eureka moment of vaccine injury didn't happen

immediately during the initial aspects of the jab roll out. I saw the odd prolonged and unusual sore arms or strange periods of severe headaches, which I myself had experienced frighteningly after the first jab where I was in bed for a couple of days with what felt like electric shocks going through my brain and into my spine. A very unusual sensation. Despite this, as discussed earlier, due to impending mandates I took the second jab and for a short while felt fine but then had a completely different experience to jab one.

It was around three to four months after this I noticed that my sharpness was slowly getting worse, I experienced cognitive dysfunction and mild memory lapses in addition to an overwhelming fatigue and moments of brain fog. I had originally put this down to the stresses of whistleblowing and my eventual blacklisting. But as things progressed with these symptoms, they could not be explained away through pressure of work or feeling particularly anxious. I remember doing a couple of podcasts where I was unable to articulate a response, and both interviews had moments that we needed to edit out as my mind kept going blank as I couldn't process the questioning. This was really quite scary, and I had started to note an uptick in others complaining in clinic and online about similar 'brain fog' experiences. This is when my research into the spike protein and possible detoxification regimes started in earnest. To cure myself of growing cognitively old before my time.

Going back to my GP clinical experience I wasn't seeing much of a groundswell in adverse events of any note that caused me concern until the third jab started to be widely administered. This is when things went from the odd reactive arthritis, prolonged lymphadenopathy and some unusual but rare neurological presentations to a more sinister flood of patients complaining of health deterioration after the jab. At the time I was working a three-day week and hence doing six clinics at a relatively small list size practice. I would see one or two sometimes three cases a day where I would hear phrases like, 'I've not been right since that jab,' or 'this all started the day after I took the covid shot,' followed up by a wider ranging and unusual set of red flags waving for all to see, for anyone with their eyes open and doing their job properly.

If I was seeing this at such frequency, surely my colleagues were seeing similar? I tried to have informal discussions about this only to be met with 'correlation doesn't equal causation,' or 'you can't prove it's the jab,' by the same purveyors of, 'Long covid.' There existed a sinister collective silence around the place as to their own experiences. It's not like I had set out to look for this warning signal, it was a repeating trend in my own clinic room of patients that had attributed their downturn in health to the jab. I couldn't ignore the evidence of my own eyes, ears, and clinical encounters.

By the way, can you prove that a single life has ever been saved by a vaccine? Obviously, you cannot. But you can prove death and injury caused by vaccines. Food for thought.

I started to see lots of young people with chest pains and heart issues, relatively young stroke victims, blood clots in a close temporal relationship to the administration of the jab. Equally these vaccinated patients kept getting covid positive results over and over again as discussed previously. Many others had reported brain fog, unusual fatigue, strange rashes, autoimmune and neurological presentations with odd constellations of symptoms as well as constantly getting ill with low grade infections which marked the era of the 'MPOX,' a condition I believe was dreamt up to explain away the sudden uptick in herpes zoster as well as other latent viral conditions like simplex and fungal infections, all signs of immunocompromise. People's immune systems were seemingly malfunctioning both over reacting (autoimmunity) and under reacting to infections as well as clear coagulopathy and links to heart issues in younger people started to gather momentum even in the world of published literature.

I was involved in a case while playing football when a player from the opposition collapsed and was in peri-arrest from a severe bout of myo-pericarditis. He had had the jab a matter of days before and had also been having un-investigated chest pains since his first. This was a nineteen-year-old fit healthy male that had absolutely no clinical indication to get jabbed other than for his work and travel plans to be fulfilled. At this time, I started looking at discharge summaries and noted upticks in all of the above but most alarmingly cardiovascular and clotting events in the young.

I mentioned earlier about my interviews with jab injured but having set up a new consultancy service outside the NHS, was quickly finding that a high number of people were booking consults having suffered the same fate as each other and sought a listening ear for advice and support. The NHS didn't recognise these common 'rarities,' no one was acknowledging the frequency of these apparent 'coincidences' let alone the pathophysiology at play. Clinics were set up as an all-encompassing approach to these odd symptom constellations that were all being stamped as 'long covid' (LC) that barely affected anyone that had refused the jab. However, it was anathema to even consider this correlation between LC and the number of jabs taken in addition to the temporal relationship to their administration.

All the stories were virtually identical. The patient had made the attribution to the jab; the patient had been dismissed as showing functional neurological disorders or mental health issues and offered a drug to this end. There was no attempt to understand the mechanisms by which these symptoms could manifest let alone the correct investigations being done. It was impossible to get a D-dimer done locally (a blood test to diagnose a blood clot), autoantibody tests were not being done (a test to show the body attacking parts of its own cell) and even in the face of positive tests such as unexplained raised inflammatory markers these patients were being gaslit that it was 'all in their head' or attributed to LC. I regularly enquire with my clients, for those 'lucky' enough to have experienced attending one of these two-year waiting list clinic slots. They offered mental health, physiotherapy, and occupational therapy support to essentially 'learn to adapt' to their new disability. No mention of pathophysiological mechanism to explain, no definitive testing being done, and certainly no spike protein attribution or detoxification protocols were proffered to the patient.

Most of what I do, that is most therapeutic to my vaccine injured clients, is to listen to them, validate them, acknowledge the gaslighting, hear and acknowledge their pain and understand the conclusions that they have come to offering empathy and sympathy. This is the crux of what we are taught as GPs.

The exam you have to pass to become a GP reinforces that GPs must consider the 'whole patient' meaning the biological, psychological, and sociological aspects of their illness in addition to acknowledging the patient's own ideas, concerns, and expectations. This gold standard training in how to get a high mark has seemingly left the building in a groundswell of complicity and a wave of ostrich head in the sand syndrome.

The ultimate case of gaslighting was happening in lock step in my profession, deny, deny, deny, ignore, hide the evidence. Young people with chest pains were not having the right heart imaging to diagnose what was a very early uptick and warning sign that things were not as safe as marketing campaigns had proclaimed. I guess if you don't seek you don't find. And even if they did find it, it was explained by the unicorn LC and the profession didn't have a clue how to manage this.

Detox protocols were labelled the foodstuff of the anti-vaxxer quackery, despite bundles of evidence of clinical improvement and efficacy. I spent a lot of time researching things to combat specifics of cellular and molecular events and pathways that were contributing to the symptoms and pathology. The answer seemingly wasn't to be 'brought to you by Pfizer' who ironically were marketing mRNA-based jabs for myocarditis and cancer in recent time to fix the consequences of their initial product. Great marketing campaign to keep the public sick as a patient cured is a customer lost.

I researched FLCCC protocols on post vaccine injury

supplements, the world council for health have published a thorough and well evidenced protocol. But one thing was clear. The detox protocols were entirely composed of cheap, healthy, non pharmaceutical natural protocols and supplements. I embarked upon a paleoketogenic diet, and a selection of supplements found to reduce spike protein burden and the improvement in my cognitive dysfunction was rapid, these included nattokinase, quercetin, natural anti-inflammatory supplements such as spirulina/chlorella and turmeric; the list went on.

Things I had never heard of let alone studied at medical school were being shown to have efficacy as antidote to the endothelial dysfunction, clotting, autoimmune/inflammatory, immune damaging carnage that the spike protein was seemingly responsible for. After all, this should have been entirely expected given the rush to market mass administration of a gene-based injection that encodes a foreign protein antigen with wide biodistribution and unknown duration of action.

Highly predictable carnage ensued and while the medical profession fell collectively silent and buried the evidence by a mass gaslighting campaign of the original 'covid dissenters' who had stepped forward from behind their disciplinary tribunals and sequential strike offs to help the victims of the damage perpetrated by the very profession cutting them loose and dismissing them. I've never seen so many people being told, after minimal cursory investigation, that their symptoms represent FND, functional neurological disorder. In other words, the doctor, in their ignorance is saying that while they can't find the cause (mainly not looking for it) it must be a figment of the patient's imagination. Faking it. Faking being unable to walk, talk, think, faking neuropathic pain, cognitive decline, complex syndromes, and constellations of symptoms. All faked by the patient.

Sickening gaslighting shown to people who trusted their doctors, nurses, and governments and 'protect the NHS' and 'protect the vulnerable' only as a consequence to suffer disability and being dismissed by the very people they trusted and who administered the shot to them. Left out in the cold alone and desperately seeking help and answers.

Not only were these victims ignored by the profession, but their stories were also being censored. Interviews like my attempt at YouTube interviews were blocked. Accounts of vocal vaccine injured were shadow banned and abused by some of the same trolls targeting me.

And then came the farce of the vaccine damage payment scheme (VDPS). Not suffice to gaslight and dismiss these victims, the paltry offer of compensation set up by the government to 'support' those with these common rarities, people attempting to access this payment were equally gaslit, dismissed, and rejected. My first YouTube interview was one that stuck in my mind. The gentleman suffered crescendo mini strokes in and around his latest covid shot on the background of good general health. On application for payment, he received a rejection as someone had dug back in his medical notes to find that he was an ex trivial smoker for a short period and had a coding of a single episode of palpitations in historic notes from decades ago. Of course, it was his smoking and a likely paroxysmal arrhythmia like the one that caused his ischaemic attacks culminating in a full stroke, absolutely nothing to do with the jab he had received days previously. This is pretty much what is happening to all applicants even in blatant cases tightly linked temporally to their disability onset. The VDPS dig through old notes and find any tenuous connection to an historic event that would add counter evidence to the jab link and henceforth rejected. How did we get to this? I will discuss the duty of candour in a later chapter.

Chapter 19 - Complaints: A Cornish dream becomes a nightmare Part 1 NHS PAG

The first realisation of my 'crimes' of whistleblowing and going against the narrative, was around January 2022. I was contacted by the NHS Performance Advisory Group local to me expressing concerns about my social media presence. They asked me to respond to the complaint that, in summary, I was affecting public confidence in the jab and spreading misinformation.

This was alarming as nowhere in the complaint were specifics of the posts or any form of rebuttal to anything that I had allegedly been 'spreading' online.

I was invited to a meeting where I faced a panel of my senior colleagues. I calmly presented my concerns and some citations to back up my position. I could tell by their facial expressions and general responses that there was no major objection to my well laid out reasoning and principles around my concerns. As a follow up, I was invited to write to them giving further detail as to my position and how I reached it which I felt was a fair question and a good opportunity to put my points across in a more academic fashion. I wrote the following letter (helped by a dear friend Teresa Tannahill) detailing my concerns, published below and referenced fully on my Substack Breaking the silence: https://drdavidcartland.substack.com/p/letter-to-the-gmc. The language is scientific so apologies in advance to my lay readers. I wanted to lay this out on a more academic basis and felt that the lack of response to this correspondence is a powerful moment in my stand. They had no rebuttal. Besides, it is clear that the non-doctor readers probably know more on the subjects discussed below than the medical covid cultists to whom it was addressed.

Dear Programme Manager

NHS England and NHS Improvement Southwest

Dear (Redacted),

Your reference: (Redacted)

You asked me to provide any information that I feel is relevant for the PAG to consider. After giving it some thought, I have decided to show you the scientific reasoning behind my stance on the COVID-19 vaccines. Where possible I have drawn on peer-reviewed papers. However, I also consider preprints, articles and anecdotal evidence to be important tools in informing opinion with reference list supplied.

When Frances Oldham Kelsey refused to authorise thalidomide for market in the United States, she had very little data to go on, but a strong feeling that something wasn't right. Thalidomide had already been approved in twenty-one countries and Kelsey found herself under considerable pressure from the manufacturers to authorise it. Despite this, she was adamant that thalidomide is not approved in the US until the manufacturers had proven that the drug was not harmful to the foetus. She also asked for clarification on a study out of the UK that documented cases of peripheral neuritis. As we now know, Frances Oldham Kelsey was vindicated when it was discovered that thalidomide crossed the placental barrier and caused several thousands of severe birth defects.

Here is my reasoning on the COVID-19 vaccines.

1) The concept behind all of the COVID-19 vaccines is that our own cells produce a fragment of the virus (the spike protein), and our immune systems produce antibodies to that fragment.

2) Unfortunately, we now know that the spike protein is not benign. It causes some of the worst symptoms of COVID-19 disease. The spike protein is a toxin, and the vaccines teach our bodies how to make it.

3) All currently available COVID-19 vaccines turn our cells into spike protein factories which flood our bodies with spike proteins in concentrations that vary from individual to individual.

4) In 2020, it was known that the virus was able to reverse transcribe into our DNA. Despite claims to the contrary by

Pfizer, their vaccine mRNA has now been shown to enter our cell nuclei. Six hours after vaccination, the vaccine begins to weave itself into our DNA, where it can produce toxic spike protein indefinitely in vitro at least.

5) Despite claims that the spike protein would be rapidly eliminated, we now know that it is produced in the body chronically. This means that the vaccinated are exposed to its toxic effects for a very long period of time.

6) A recent autopsy study revealed that deceased, vaccinated patients had severe pathology to their vital organs caused by the vaccine. The pathologist determined that the vaccine had caused 14 out of 15 of the deaths. Families in most countries are finding it extremely difficult to convince coroners to conduct full autopsies. The net effect is to obscure evidence of a causal link between the vaccines and death, especially since most of the bodies are soon cremated. I am not alone in finding this concerning.

7) The spike protein seriously damages the lining of our heart and blood vessels. It causes microscopic blood clots that prevent the lungs from adequately supplying oxygen to our body. In susceptible individuals, these micro blood clots clump together. This can lead to fatal COVID-19 in all ages, as well as the myocarditis, strokes and heart attacks seen in younger people who receive the vaccine.

The majority of these vaccine injuries occur within 48 hours of injection. Around 80% of vaccine deaths occur in the first 14 days after vaccination. For many, the micro clotting may manifest as tiredness and lethargy as the blood loses its oxygen carrying capacity, but each booster shot increases the risk of larger clots forming. Colleagues have remarked on elevated D-dimer findings in vaccinated individuals, which supports this explanation. The fact that no formal studies are being carried out to investigate this makes me concerned.

8) Analysis of the US VAERS database reveals that a small number of batches from each manufacturer were exceptionally toxic, leading to an extremely high number of strokes, heart attacks, severe reactions and deaths when distributed. Pfizer, Moderna and J&J's products all showed excessive batch-to-batch variation, with fluctuations more than 10 times greater than flu vaccines. At best, this indicates grossly negligent non-compliant manufacturing practice. At worst, malfeasance. The problematic vaccines should have been recalled.

9) The spike protein doesn't just embed itself into our DNA. It also soaks up the BRCA1 and 53BP1 proteins that are essential to patch up our DNA when it gets damaged by a number of environmental factors, e.g., sunlight and pollution. It is clear that this can only cause degradation of our DNA which will inevitably lead to a number of unpleasant syndromes, including rapid ageing, shortened lifespan, neurological disorders, and cancers.

10) The vaccines also force our white blood cells (lymphocytes) to express spike proteins on their surface. This makes them look suspiciously like viruses. Since our CD8 killer T cell lymphocytes are programmed to destroy viruses, they immediately start killing themselves and each other with catastrophic consequences. Our immune system begins to die off. This is likely to result in more frequent and more serious bacterial and viral infections. Some doctors are already describing cases of 'V-AIDS,' in which patients have become seriously immune compromised following vaccination. As a physician, I would be more reassured if, instead of the plethora of 'fact checking' articles debunking the existence of V-AIDS, some investigational studies were carried out. Until we get more data, I believe that the only rational approach is to take this worrying anecdotal evidence as a safety signal.

11) Lymphocytes are also responsible for keeping cancers in check. Doctors worldwide are now reporting a huge rise in sudden, aggressive 'turbo-cancers' in the vaccinated. Patients in remission are experiencing a resurgence of their cancers after getting vaccinated. In the absence of any official effort to investigate this phenomenon, the only rational response for any physician is to be informed by the mounting anecdotal evidence and treat it as a safety signal.

12) Part of the spike protein, the S1 subunit, appears to have been designed to break off at a point called the

‘PRRA polybasic furin cleavage site.’ This feature makes it exceptionally good at infecting humans, which does seem to support the gain-of-function lab origin theory. The PRRA sequence has never been seen in any naturally occurring coronavirus, but it does appear in a MODeRNA patent dating from 2016. The chance of a virus naturally mutating to match the sequence in the MODeRNA patent has been calculated as three trillion to one.

13) After it breaks off, the S1 subunit is capable of travelling through our spinal cord and up into our brain. We now know that the S1 subunit contains a prion domain that can cause the types of damage to brain cells associated with CJD and dementia. The prion domain in the SARS-CoV-2 spike protein is not present in other coronaviruses.

14) Humanised mice challenged with the spike protein developed spongiform encephalopathy (CJD), colloquially known as ‘mad cow disease.’ There have been many reports of sudden onset CJD in the vaccinated. Coincidentally, there have also been recent reports of BSE in cattle, including one case on a farm in the UK. Interestingly, the EU has just re-legalised the feeding of animal carcasses to livestock. Oddly, a disproportionate amount of media coverage has been given to this man’s strange culinary habits. I sincerely hope I’m wrong, but I would not be at all surprised to see CJD in the headlines again at some point.

15) Macaque monkeys tested with the spike protein

developed Lewy bodies. Lewy body dementia is rapidly progressive, with patients typically dying 5-7 years after diagnosis. There are many reports of sudden onset dementia in people who have had the vaccine.

16) The lipid nanoparticles that envelop the vaccine mRNA are themselves toxic. A Pfizer report to the Japanese government showed that these nanoparticles rapidly accumulated in the spleen, liver, and ovaries. As one colleague recently pointed out, technology is being developed to target LNPs at specific cells in the body. Our spleen, liver and ovaries are essentially our immune, detoxification and reproductive systems. Again, I find this deeply concerning.

17) Both Pfizer and MODeRNA reports to the EMA showed that infertility doubled in vaccinated rats compared to the control group. MODeRNA noted that pregnant vaccinated female rats lost fur, suffered swollen hind legs, and had limited use of their hind legs. However, MODeRNA didn't consider these effects to be 'adverse.' They also noted skeletal abnormalities in pups born to vaccinated rats.

18) The vaccine was designed for the original Wuhan virus. The virus has mutated many times since and the vaccines are now more than two years out of date. The antibodies they produce are highly tailored to the spike protein and are only produced in the blood. Natural infection produces antibodies in the mucosal membranes, where

they are greatly needed. The latest UKHSA data shows that the current crop of COVID-19 vaccines offer little or no protection. As a doctor, I am concerned at the push to 'boost' with these out-of-date injectables, given that they are adding very little benefit, yet increasing exposure to risk.

19) While researching, I discovered that several experimental coronavirus vaccines were abandoned after it was discovered that they caused Antibody Dependent Enhancement (ADE). The animals in the trials died. The 2020 trials did not check for ADE reactions in humans. There is evidence that we're seeing ADE with the COVID-19 vaccines. The so-called 'breakthrough cases' being observed in the vaccinated may be a form of ADE.

20) The vaccine doesn't prevent the vaccinated from catching or spreading the virus. Instead, it appears to limit the severity of the disease by suppressing our immune response. This allows the virus to persist for an unnaturally long time, which creates the perfect environment for it to mutate and evade the vaccines. In effect, every vaccinated person becomes an individual gain-of-function laboratory, harbouring and incubating the virus, potentially causing more deadly vaccine-resistant variants. Several eminent scientists including Dr Geert Vanden Bossche and Nobel laureate virologist Professor Luc Montagnier have warned that vaccinating indiscriminately in the midst of a pandemic

is not a scientifically sound policy. The data out of Israel and the UKHSE appears to be proving them correct.

As you can see, I have a strong argument for my cautious stance on the COVID-19 vaccines. I'm sure you can imagine, it was quite a shock to discover that the vaccine spike protein could cause the above side effects. I initially felt upset that I had been misled as to the safety and efficacy of the vaccine. If a friend, family member or patient expressed similar views, I would have to empathise with them. I know exactly how it feels to realise that you've been injected with an irreversible gene therapy that could potentially cause you acute and long-term adverse health problems that you would never have consented to had you been properly informed.

My learning from the above experience is that I recognise the importance of being empathic and open-minded to what my patients tell me. Having had the unnerving experience of finding that magnets stick to my injection site, despite there being no magnetic API, adjuvants or excipients listed on the vaccine ingredients, I have valid and pressing questions for which I have to date received no satisfactory answers. I am not the only person seeking these answers. In a 15th March 2022 letter to a UK Column journalist the RCGP admitted that, "The reason the GP cannot give you long-term information on side effects or the exact ingredient of a vaccine is because that information is not available to them."

Going forward, I will be following the vaccine safety and efficacy data with great interest. I would also like to participate in any studies that might shed a light on the long-term effects of the vaccines. Due diligence — albeit late — is better than none at all. The WHO state that 67% of humanity has already received at least one injection. They expect to reach 70% by midsummer and 75% by the end of 2022. If the COVID-19 vaccines turn out to be another 'thalidomide,' there will be no going back.

I look forward to discussing these points with you. I am more than happy to look at additional safety data as it becomes available, so please send anything on that you think I should take a look at. Meanwhile, I will continue to maintain a cautious approach.

Yours sincerely,

Dr David Cartland

Chillingly, but no longer a surprise, I never had a reply to this letter and have never heard from the NHS PAG since, which pretty much says it all. Evidence against and rebuttal of my position could have easily shut me up and put me back in my box but instead of which we are met with a consistent paucity of debate.

Chapter 20 - Complaints: A Cornish dream becomes a nightmare Part 2 GMC

In and around January 2023, I entered a phase of wanting to call out public misinformation, government scientists and decision makers, so I decided to use my X platform to highlight some of the blatant inaccuracies, premature claims of safety and efficacy and other scientific breaches of logic. I set up the rogues' gallery. Daily I would post one of the high-profile public health doctors and scientists making unsupported claims of 'safe and effective,' alongside their image, a written version of their claim and often a video or link to them saying it. I did this to shine a light on what was being promoted on large audience platforms and mainstream media rather than individual targetting. Accountability was important especially if people had believed these 'trustworthy influencers' and government officials.

Around the time of doing this I received my first few notifications that a GMC complaint was being looked into further, having received prior to this several vexatious and frivolous attempts by my trolls for 'spreading misinformation' to report me to the regulator. Interestingly, I recently did a subject access request to the GMC to see how many complaints I have had and when they started.

The result of this enquiry was fascinating.

On the background of an exemplary academic and clinical record, some two months after my first podcast and public spotlighting of concerns, was the first flurry going back to February 2022 and nothing recorded previously to the GMC. To date I am told, I have had 243 complaints about me almost exclusively around ironically, 'misinformation' and mainly from anonymous reporters. Some that did have their identity came from within the medical profession without many exceptions. When these complaints are closed (238 at time of writing), my local responsible officer (who incidentally has never contacted me let alone spoken with me to offer remedial advice or warnings on conduct) is notified of the baseless and closed reports. I understand a number of meetings were held about me locally, led by my responsible officer of which I was not informed or invited to give self-defence. The outcome of these meetings was that I must reflect on these events as part of my annual appraisal and portfolio content.

Back to January 2023, one of the individuals featured in my rogues' gallery had written to the GMC to report me, saying that my post had been harassing in nature and they felt I had incited my followers to attack, and led to mental health symptoms. Shortly after, a similar complaint was filed by the alleged ringleader of my online trolls, closely followed by another doctor that I had called out publicly.

Suddenly, the GMC had a line of attack. Dr Cartland is a vile bully, harasser, and inciter of a frenzy of crazy anti-vaxxers.

One of the sickening ironies is that one of them, a ringleader, had harassed me for the best part of two years playing DARVO tactics (Deny, Attack, Reverse Victim and Offender) on me while hoodwinking the GMC. Anyone who defended me or expressed any 'anti-vax' views would face a pile-on by them and their coordinated group, facing a torrent of ad hominem abuse. Everyone was aware of this character's online antics. Call me cynical but I noted that the GMC, similar to the NHS PAG, did not seem willing to take on science and rebut the 'anti-vaxxer' rhetoric head on, instead targeting dissenters with other mudslinging accusations hoping some would stick. Ironically, the 238 complaints had been thrown out on the overarching principle of my freedom of speech and expression needing to be honoured. Sweet irony, and not much compensation for those like Dr Sam White and Mr Mohammed Adil who were struck off for doing exactly that, exercising freedom of speech and scientific expression. Dr Daniel Armstrong has been struck off in the interim for stating his concerns re exaggerated safety and efficacy claims of the covid jab.

At the time of being notified of these three complaints, the GMC have a process leading up to a decision by case examiners as to whether to proceed with the case. As you may or may not be aware, a GMC investigation is highly stressful and should be conducted swiftly. Thirty months after the initial notification the investigation was still ongoing with my inbox continuously full of further correspondence of other rejected complaints or the next stage of the investigation. As an observation the GMC appear to love sending such emails on a Friday evening or just before a bank holiday weekend. Wonder why they would do that? Is this part of their attritional tactics towards dissenters? I am sure you will have worked that one out.

One of the things the GMC insisted on was that they wished to send a letter out to the places that I worked as part of their protocol. I resisted this as a) I was not guilty of the claims being made b) I had not been given a chance to rebut or defend the claims and c) the reputational damage this would do to me and my career locally. I was harangued for weeks to complete this 'work detail form' eventually being warned that if I didn't submit to naming my practice, that I was apparently breaking the law and this would be added to the charge sheet.

I relented and the letter was sent to my local six regular locum practices where I enjoyed my experiences and always received good feedback. I managed to get my hands on one of these letters as one practice was astounded at the contents and the principle of them; sending out an 'FYI' letter based on undefended accusations and that I was subject to a 'fitness to practice' tribunal. They also make a clear attempt at the end of the letter for the practices to offer to the GMC any concerns (not positives) that they may have after being notified of my alleged crimes... Digging for dirt!

Following on from this and almost instantly, the shifts dried up, offers of work became increasingly sparse. Understandable given their receipt of a letter saying I am subject to fitness to practice for harassment and incitement. Unfortunately for me, I work in a remote area where there is a very strong chain of gossip and between the undercover meetings that had been going on behind my back and these letters going out, a lock step blacklisting of Dr Cartland ensued that has continued to the point of writing this. Cancel culture at its finest. Interestingly, rather than reasoned debate or adult conversation, when I email practices for locum work or to apply for a salaried role, it is very rare to receive a reply let alone a rejection letter. By the way, I suspect they take the position that if they don't respond, they don't have to justify their decision which would of course be based on politics not medicine.

I recently worked two short stints at local Cornwall practices who ironically had staffing issues due to multiple doctors being off sick. The feedback while I was there was excellent, one manager saying patient feedback had been 'exemplary' and 'unprecedented' reminding me that there was a salaried role coming. This also happened at the other surgery. After three days at each place, I did not hear back from either of the two practices about my job application that they themselves advised me to put in and I am reliably informed by my trolls that two individuals from X had kindly contacted the practices to inform them of my 'charge sheet' and impending tribunal reconvening.

In a moment of truce, I recently wrote an email to all surgeries in Cornwall offering a presentation, at my expense, to discuss my experience of vaccine injury and research around detoxification. I never received a response, not one. Likewise, two surgeries had sent two of my friends a text (a now common occurrence) saying that due to staff sickness there were no doctors on site. I wrote to these two surgeries with read receipt emails and never received a response. So much for patient safety. There was a recent critical incident at the main hospital in Cornwall with massive waiting times and compromised patient care as a consequence. I wrote an email to the consultant body and the executive board to offer a 12-hour pro bono shift given my dual training in A+E and GP in addition to a background in A+E clinical navigation. As became the pattern, I never received a response.

Moving swiftly on to GMC proceedings and careful not to get into further trouble, since the original complaint I had two further complaints shoehorned into the October

2024 MPTS tribunal proceedings. One for calling out an understaffed local surgery who rejected my application to work there based on a google search. Given they had recently won an award for inclusivity, equality, and diversity, I felt that this was not being demonstrated. My post about this surgery led to a 'flurry of hate mail' being sent to the surgery. Three letters in fact. Two by science professionals and one from a retired local professional man each referring the practice manager to the Pfizer trials! Hardly hate mail. This had been misrepresented by the GMC as my incitement of targeting the practice manager personally, which was absolutely not the intention of this post which ironically was about patient safety concerns about understaffing and provision of a safe service highlighting this locally. I had been careful to redact the practice manager's email address before posting the letter, but this was subsequently altered to reveal her details and given to the practice by an anonymous individual supplied in a brown envelope. Nothing to see here!

The final case is about my alleged offer to 'provide an individual with a covid exemption' and to 'create a medical record containing information which he knew may be false and his actions were dishonest.' The truth of the matter is more complex as I am sure you have already guessed. I was contacted by a man claiming to be a locum rep finding work for me in the area. A few months after this enquiry on WhatsApp, 'Mark' (who turned out not to exist) from

‘Atlantis Medical Group’ (which didn’t exist) who claimed to have heard that I was able to offer vaccine exemption letters and that he had an urgent ‘make or break’ career interview abroad and needed one to travel. Prior to this I had done around five of these exemptions based on two principles; Bodily autonomy and ‘Primum non nocere’ first do no harm, given my research leading to the already stated ‘unsafe and defective’ conclusion. This was not an advertised service, nor did I charge for this. All of the exemptions were done for people in dire situations, one being a grandmother who contacted me asking for a letter to exempt her because her daughter and first-born granddaughter were critically ill in a New York hospital after her daughter had unexpectedly and prematurely given birth. She asked her GP for an exemption in order to travel; they had declined and gave no reason for doing so. I provided a letter on the grounds previously discussed and she travelled to the side of her family and sent me a selfie of her with daughter and granddaughter fully recovered along with a heartfelt thank you. The other exemptions were of similar ilk.

Back to ‘Mark’ who was offered a consult to discuss further and when he enquired regarding cost, I said it was on a pro bono basis, aka free and without charge. I won’t repeat the final message, but it involved an offer to sell me a caravan for when I lose everything followed by a number of ‘c’ based profanities. This had obviously been a set-up, an entrapment. Something that despite the legal system not being allowed to be used as evidence, the GMC proudly claim they can use including ‘hearsay’ in tribunal proceedings. How did the GMC get involved I hear you cry? Well Mark had the foresight to send the WhatsApp exchange (minus his final abusive message) to the GMC. Not directly but via two of the current complainants who submitted almost identical submissions to the GMC within just over an hour of each other on the same day. Nothing coordinated, move along.

Anyway, that was the case against me in its entirety.

In October 2024, some eighteen months after the decision was made for fitness to practice, the hearing began in earnest. The MPTS are a panel who listen to the evidence presented by me (the registrant) and then the GMC case against me. The panel is made up of a lay member, a medic, and a legally qualified chair. Without naming names the GMC had brought in the big guns for this ‘he said she said’ twitter spat, a gentleman whose day job is in terrorism and homicide barrister work. The legally qualified chair was none other than the Chief Prosecutor for the Crown Prosecution Service, London. Both taking initially around four weeks out (became six weeks) to adjudicate on this case. I was represented by a barrister and the case began. Well, I say ‘case’ politely and loosely as what ensued over the following days was farcical and unprecedented.

The initial discussions were around whether the case should be heard publicly or privately to protect reputations of myself and the complainants and provide a safe space to discuss issues transparently and openly. We proceeded to read out an email stating to the GMC that the (GMC redacted) author was a member of a 'network of scientists and doctors from the X platform' (available to view publicly on my X feed) and that I had been allocated as a 'TARGET of their group.' In black and white, I am a target of a group of X crusaders against 'misinformation.'

We then proceeded to read out a crime number and a number of threats made to me by X trolls, in addition to other evidence of doxing, abuse, harassment, hacking and cyberstalking to support the application of privacy. The GMC had already allowed a number of my most prominent trolls to watch this hearing, having applied for a proverbial 'golden ticket' to watch proceedings, many trolls openly boasting that they had booked time off work to witness my 'glorious demise.' The GMC were made aware of this, and no action was taken to remove this intimidatory audience. The panel adjourned for the day and returned with their verdict. To cut a long story short, the outcomes of three days toing and froing on administrative issues before we began proceedings formally, they had decided: the complainants were to be protected by an anonymity order instead being referred to as Drs A/B/C and that they were to be treated as vulnerable witnesses. I didn't see or hear any evidence to support this decision hot off reading out threats to harm me including death threats by other third party trolls. The panel decided that two complainants would also have their evidence heard in private. Dr David Cartland however was deemed to be heard in public, not a vulnerable witness, no such anonymity order granted and that my testimony would be beamed out publicly including to my online trolls.

The next item on the agenda was the admissibility and relevance of my evidence and witnesses. Essentially, I had done sentence by sentence rebuttal with evidence supporting my position to every single statement against me, a portfolio of glowing testimonies and references to my character. I also included a couple of examples of double standards from the GMC e.g. refusing to investigate a death threat and numerous abusive posts from a medic/troll, in addition to a bundle of work I had compiled showing the interconnections to the trolls, their interactions and more interestingly connection to two of the GMC medical complainants. Thorough and highly relevant to context and my defence. I did a case study of an anonymous troll, one of my worst and fixated, showing connections as stated. Finally, and despite being capable of providing 100s of witnesses that it was in fact I that was the victim of a mass harassment campaign not the perpetrator, I decided to streamline it to six witnesses, all highly relevant.

Anyway, you know what happened next, I'm sure. The

hearing was adjourned, and it was somehow decided that three of my witness statements were to be ruled inadmissible and irrelevant and the three remaining were redacted to the point the content was not relevant to my defence. All references to the campaign of harassment towards me and evidence demonstrating a culture of online bullying by trolls towards me, as at least a mitigating factor, were blocked from being heard at the hearing. The GMC barrister commented that the 'vast majority' of my own defence bundle was equally irrelevant and inadmissible. My own words, analysis, and evidence of a mass coordinated campaign against me was irrelevant to the hearing taking context fully out of the proceedings. We applied at this juncture for a strike out of the case given the abuse of the human rights act for the right of free speech and a fair trial and flouting of legal process. My barrister has never heard such a thing as a defendant not being allowed his witnesses to speak. I'm sure you can guess the outcome. Rejected.

So, without witness statements, most of my own statement inadmissible and the protection of the 'victims' paramount, while I am beamed out publicly to my trolls with no such protection, we start the proceedings. The tribunal was a virtual hearing due to logistics of getting everyone in a room for four weeks in Manchester.

Essentially my role was to sit and listen and I was not once permitted to speak unless through my barrister. A virtual house arrest ensued whereby I had to be available at the click of a finger as the tribunal, at snail's pace, pontificate over nothingness while the meter was running on my legal costs. I would like nothing more than to say in three weeks a lot of ground was covered but this would be a gross overexaggeration. In summary, the case was opened against me, I sat for two hours listening to the terrorist barrister misinterpret, remove context from, blatantly misdirect the audience as to the case against me, making me out to be a vile, criminal harasser sitting in my basement abusing innocent doctors who were simply 'following national and international guidelines.' It was a sickening thing to watch, always wondering has this man actually read the witness bundle he tried to have thrown out in its almost entirety. If he had he would have clearly realised, for example, that a parody account I unwisely set up about one of the doctors was in retaliation of HIS abuse of me over two years, in addition to himself and fellow trolls setting up numerous parody accounts in my name 'Dr Cartload of bulls£$t' a particular favourite. This was conveniently missed in the details of his 'Dr Cartland is a bully' monologue.

Over the next couple of days, I watched on as we cross-examined two of the witnesses. It was one of the most surreal experiences of my life watching two alleged professionals give very different accounts of what actually happened. My summary was that their performances were an embarrassment and anyone watching (of course that would be no one as the ruling to have privacy in their parts of the hearing stood) would have spotted the inaccuracies and stumbling false narrative that my barrister managed to dissect and shine a light on. I felt embarrassed at one point for one of the complainants as they fumbled through their lines, becoming animated and eventually threatening to sue my barrister for slander! At another point the legal chair had to step in to deescalate the ranting. A complete court jester of a cross-examination witness. Any neutral and sane panel would have thrown these two witnesses out for their 'performances,' but no. This is a witch-hunt for Dr Cartland's medical licence.

Proceedings came to a halt at the end of a painful two weeks and things had escalated for me in terms of a prolonged period of insomnia, constant hypervigilance due to stress anxiety and low mood reactive to the events combined with the duration that this has been going on. When we adjourned on Friday evening, I felt overwhelmed by emotion and had tried to distract myself with a film and a beer. The anxiety intensified as I played back in my mind the events of the fortnight. Literally nothing was going for me, every decision against me, often flouting legal process and human rights, and it was that night that I woke in the small hours ready to end it all. I just didn't see a way out. This was clearly orchestrated, and my speaking out had upset some powerful people in the world of big pharma, and my take down was to be collateral damage to keep up the pretence of 'safe and effective' by reputational damage of a strike off to invalidate my public exposure of what was really going on.

It could be argued that the GMC have a track record of pushing people to suicide. According to a BMJ article, a survey reported that one out of three doctors had suicidal thoughts during a GMC 'process' and between January 2018 and December 2020, a three-year period, twenty-nine doctors died while under GMC investigation. It is even more likely when you are blacklisted from working due to their system of informing workplaces and treating registrants as guilty until proven innocent while taking the words and accusations of internet trolls against a professional doctor.

I don't want to discuss the finer details of that night other than to say I was at the lowest I have ever felt in my life, the stresses building up like a tsunami, the chances of a positive outcome unlikely and the pressure insurmountable. The GMC and the trolls are responsible for my night at the cliff edge. The surgeries blacklisting me led to my night at the cliff edge. The ex-colleagues and friends shunning me led me to my night at the cliff edge. The GMC complainants led me to my night at the cliff edge. In contrast, my patients supported me throughout and for that I will always be grateful. Bear in mind in all of this there was not one complaint from a patient of mine; in the end that is all that matters to me.

Over the next few days, the police had triggered a crisis team assessment as part of the follow up of that night, which I attended. I contacted a private forensic psychiatrist for an independent non-NHS assessment. It took the GMC four further days, when presented with the harrowing events of that evening, to finally adjourn the case until June 11th, 2025, some seven months later, to prolong the torture into a third year. The GMC clearly wanted to continue in my absence and the best they could offer otherwise was a more regular 'comfort break' turning off my camera and making proceedings private going forward.

Since I de-instructed my barrister, I had written to the GMC to advise them that I would not be attending what amounted to little more than a kangaroo court in person and that I would rely entirely on my full written submissions which I hoped they would be inclined to read and digest the contents of.

I hope and pray for a good outcome but given proceedings thus far I very much doubt that. Suspending me three years after the event given there has been no interim measures or restrictions on my practice would be a very strange decision and hard to justify in addition to a strike off given a crime hasn't been committed and no patient has been harmed with any clinical negligence. What is the real remit of this registered charity regulator which exists apparently to protect patients, yet their neutrality is brought into question as soon as you look at the regulator's investment portfolio which at the time of writing includes Merck and Astra Zeneca.

Interestingly the GMC insisted that I undergo psychiatric assessment by a GMC appointed psychiatrist on top of my own NHS and private psychiatric opinions which described the harrowing impact of the investigation and recent events. Likewise, the GMC continue to contact me with further frivolous and baseless complaints from all and sundry and have advised they are lining up the next fitness to practice; keeping me busy with hearings for the next few years. The latest four complaints are sickeningly vexatious and malicious and even the GMC investigator didn't really know what the charges were including a further example of targeted entrapment by a troll who offered to unredact the name of the author of the email saying they were targeting me on X only for him to be a stooge and sent the email exchange to the GMC. 'Dr Cartland asked me to unredact the redaction of a complainant wishing to remain anonymous which is dishonest and unethical reads the charge sheet'! I've since had two further complaints for sharing holistic cancer protocols despite including a strict and thorough disclaimer to seek your own doctors' advice and do your research. I clearly stated these were FYI posts by third party naturopaths. The witch-hunt continues in earnest. The tribunal was set to reconvene on 11th June 2025.

Chapter 21 - Career ending kangaroo court

I have written this chapter at a different time to the rest of the book which was pre-outcome of the tribunal. To say that the last few weeks have been a harrowing combination of emotional anger, distress, and injustice, would be an understatement. Since the conclusion of the hearing nearly three weeks ago, I have been struggling to come back to writing the concluding chapter of this book. I have opened my laptop several times only to well up with tears, PTSD strikes, and attempt abandoned. Writers block. As I type this now, I have tears in my eyes. My whole life's work has come to an end, my dreams, and aspirations for the remaining years of being a doctor cruelly and coldly snubbed out.

The adjournment of the tribunal in October due to the serious impact on my mental health had led to a dual psychiatric assessment instigated by the GMC as a condition for the adjournment, which I reluctantly agreed to. My own psychiatrist, who had been excellent and understanding, provided an initial assessment, and subsequently a GMC psychiatrist from none other than Imperial College London, followed with a dual assessment. To be fair to the professor, he was thorough, showed empathy and understood my position and its impact speaking against the

NHS consensus and its obvious repercussions. Long story short they both agreed that I was sane (thankfully) but that the reconvening tribunal may be triggering and so best to proceed in private forum with cameras off and no trolls present to goad and harass. I agreed to returning in person, on this recommendation.

In the interim period, my barrister had reconnected with me asking if he could continue to represent me, as he felt that what had occurred was deeply unjust and wanted to rectify this. I reinstructed the barrister and informed the GMC of my own concerns of the troll behaviour towards me escalating online as the tribunal recommencement neared. I feared a recurrence of mental health deterioration so decided to allow the barrister to represent me in my absence, having prepared a statement to be read out and that I would rely wholly on my comprehensive witness rebuttal bundle.

Just prior to the start of the proceedings, I had been working for five to six weeks at a surgery many miles from my home address and was settled back into a GP locum role. The feedback from staff and patients had been excellent and even described as 'unprecedented.' I even had the odd request for a selfie and an autograph or two from patients who recognised me. I had been placed under no interim measures, no conditions of practice, and deemed free and safe to practice during this 2.5-year juncture of investigation. What was particularly pleasing and reflective of the trust in me, was that the surgery trusted me enough to run what is called the 'duty' doctor shifts, essentially in charge of the day and strategizing how to deal with the patient demand, triage and distribution of workload.

On the day of my tribunal restarting, the surgery received two menacing phone calls (one at the exact time of recommencement 09.30) from two males who refused to give their names but who advised that they had patient safety concerns due to my working there. These people had somehow discovered my location some two hours from home. I had been open and honest in my dialogue of the tribunal about which I had numerous frank discussions. Many of the clinical staff found it hard to disagree with my concerns and agreed that they had seen similar but were afraid to speak out as it was a poisoned chalice and career ending whistle blowing as evidenced by the outcome of my fellow dissenters GMC hearings leading to universal strike off. Despite this, the surgery had continued to give the latest spring booster (in the summer), and I had started to meet folk that had had eight or nine jabs and one who had received ten! The two anonymous callers threatened to contact journalists and the CQC about my locum work there and by 7pm on that evening and without discussion, just as I had got my confidence back and started to get my head above water financially, ping, an email dropped in relieving me of my duties, without discussion or reasoning.

In the same week, the abuse online stepped up, together with even more attempts to derail my life such as contacting the police, child services, neighbours, the FA (my hobby of grassroots refereeing), and locals. My personal email was being used to open up bondage and pornography websites in my name and AI was being used to create derogatory memes and images of me, in order to harass and defame me. All of this had been passed on to the GMC as a mitigation factor that I had been under siege from my own personal targeting by trolls, but not once had this been considered or referred to in any part of the tribunal, in fact it was actively dismissed as irrelevant and inadmissible.

I stayed away as described above, as the third and fourth witnesses were cross-examined. Filled with misrepresentation, misattribution, and misinterpretation the 'Dr Cartland is a bully' narrative stepped up again in earnest. Interestingly, one of the first things the tribunal stated was that no negative inferences would be made from my non-attendance given my representation and copious rebuttal documentation, hardly non-engagement. Despite this, a multitude of times negative inferences were repeatedly made that I hadn't 'been bothered' to turn up according to the GMC barrister. During the closing statements of the GMC barrister, he continued to misdirect and misrepresent to the public audience (due to my non-attendance the tribunal had defaulted to the 'public for Cartland, private for witnesses' ruling).

At this juncture it's important to state that before the recommencement of proceedings, the GMC had been actively monitoring my social media account gathering further 'evidence' adding to the witch-hunt feel of proceedings against me.

Back to proceedings and the GMC counsel's continued misrepresentation and context stripping of events. My barrister spoke about my lack of patient complaints and lack of clinical negligence, my excellent reputation and relationship with patients and colleagues, and that online social media disagreements were bidirectional, not unilateral. I have been open and honest to what I am guilty ie calling out misinformation and inaccuracy of medical and scientific colleagues (re-labelled harassment by the GMC) and calling out bullies.

Likewise, I was being accused of dishonesty for an act of offering to consult with a troll who requested an exemption for which I offered to consult pro bono (this complaint was put to the GMC via two of the three doctor complainants on the same day as mentioned earlier – a fact dismissed as coincidence), and frustratingly posting an email received rejecting my application due to the GMC instigated local blacklisting for being an 'anti-vaxxer.' That was it, three bidirectional online spats (not considered relevant by the panel) hence a rather confusing definition of harassment being utilised. So, without witness statements being heard, with my own mitigations and analysis of collaboration and coordination being dismissed, my own bundle being disregarded, and context stripped from all bi-directional online exchanges, the GMC adjourned to consider the facts and whether they were proven. Seventeen in total.

The panel returned the next day to find that all counts were proven. Did Dr Cartland say this in a post… he did. Did Dr Cartland refer to this Dr as that… he did. All context and reasoning removed. I was then assessed as to whether this constituted misconduct and whether, if this were the case, that this guilt after three years of unrestricted practice impaired fitness to practice, bearing in mind I had been working independently three weeks prior. Severe misconduct charge… GUILTY. Fitness to practice impaired due to this… GUILTY!

At this point, I could barely believe what was being communicated to me daily by my barrister and a couple of journalists that had been in the audience. There came a short adjournment to consider what the disciplinary action would be. Even at this point a strike off for something like this was far from my mind and I had resigned to a short suspension with some remedial training in social media and a chance to redeem myself. How mistaken I had been. The decision was made to ERASE my name from the medical register with immediate suspension pending appeal (lasting a month). Struck off! I had been given the exact sanction as Harold Shipman, while the GMC were presiding over a number of dubious recent suspensions (not erasure) of doctors that had been involved in sex offences, rape, storing images related to paedophilia, criminal acts and drug and alcohol issues. Where was the proportionality in this case? Where was the chance to retrain and be offered support to help with any issues raised, especially when there was no clinical aspect to this case? In the interests of transparency, the gory details of all this, obviously without my own testimony, are available on the MPTS website.

As an aside, interestingly, the local Cornish rag being egged along by a small number of my trolls had decided to publish this outcome days before it had even been announced, demonstrating potential leaks direct from the GMC, and proceeded to post derogatory content about me eight times in two weeks which was unprecedented for this outlet, and at a point when I was at an all-time low. Even a hit piece that I was in some way homophobic for opposing trans-GENDER concerns in sport and child safeguarding! They had trawled my social media for posts as far back as 2-3 years prior and wrote a hit piece entitled 'Inside the warped mind of dodgy Cornwall Doctor and his homophobia.' Seemingly the next phase of deconstructing my life is local reputational damage infliction.

The attacks on me online, the frothing of the trolls, and their jubilation at the result, continued unabated; their outpouring of harassment, alongside the GMC outcome, drove me to a challenging place.

I am so incredibly sad, angry, and grieving that I may never be able to work as a registered doctor again - pending a likely fruitless and expensive appeal within a regime that has already declared itself more likely to bow to the nonsense of pharma funded trolls, while those doctors who strive to first do no harm, risk being struck off.

My whistle blow was wholly and completely for my patient-centred, ethical, and evidence-based concerns, sacrificing everything for the Hippocratic oath that I applied daily, only for the injustices of what others described as a 'kangaroo court,' its warped decision making, and unprecedented bias.

I had been found guilty of harassment, despite no criminal convictions or even a police interview. The Tribunal only has to find 'on balance of probability' so if four witnesses say I harassed them while I say I didn't, that seems to be good enough for a conviction and permanent ERASURE. No 'beyond reasonable doubt' required to end a career even with an acknowledgement that there was not a single patient complaint and this was only about 'public trust in the profession.' Some might ask why should the public trust a profession whose governing body would strike off a doctor for a social media spat while only suspending doctors proven of rape of a patient, another storing pornographic images of children, and one leaving an operating theatre during an operation to have sex with a nurse. The list goes on.

I had to preside over my reputation being irreversibly tarnished and defamed as I looked on while I was unceremoniously transfigured into a bullying, harassing, dishonest person, guilty of such serious misconduct that I was deemed a threat to patient safety and bringing the profession into disrepute.

My entire life and training has been geared towards one goal, helping my patients, something I will be unable to proceed with, for now, in the conventional sense. As I commit these words to the pages of this book, I am still in the shock and disbelief stage of my grief, but I have begun to regroup, and while I have no doubt there will be difficulties, I know I will be stronger for this time.

I thank all of my supporters for their kind support and messages that have helped the healing process commence and look forward to coming back stronger after a short period of recovery.

Chapter 22 - Duty of candour... a reminder

On the 1st anniversary of the GMC's ironic revision of their Good Medical Practice (GMP) guidelines, I was sent an email article to bring fresh attention to its contents. For once I agree with the GMC. Every single Doctor practicing under the remit of this charity regulator needs to revise this as part of their evening bedtime reading routine and take note of its contents pertaining to an overview of practice values, standards, and ethical principles.

The last five years has begged many questions from the public towards medical NHS staff given some very unusual pandemic policy, totalitarian decisions, unprecedented mandating of novel 'vaccines,' ethical breaches, and evidence-based safety concerns. There has categorically been a negative shift in public trust in the profession during and following on from the pandemic years. Lower levels of trust are associated with lesser likelihood of pursuing medical advice. Consequently, restoring trust represents a public health imperative.

The good medical practice guidelines are very clear regarding a number of key areas considered as follows and detailed further in turn in the article. Making the care of patients the first concern, providing a good standard of practice and care, and working within competence, working in partnership with patients and supporting them to make informed decisions about their care, treating colleagues with respect and helping to create an environment that is compassionate, supportive and fair, acting with honesty and integrity and being open if things go wrong and finally protecting and promoting the health of patients and the public.

In view of the above facts all being equally important, I draw you to the following points re: 'being open if things go wrong,' 'honesty and integrity,' and 'protecting and promoting health,' when treating patients and the public.

There appears to be a divide between the medical profession's acknowledgement of the glaring issues with pandemic medical and governmental policy and overreach, lack of provision of COVID treatment advice, ethical breaches regarding matters such as informed consent and bodily autonomy, in addition to a cultist type 'religion' of 'safe and effective' at all costs. This despite mounting unequivocal evidence that brings 'safe and effective' not only into question but also that COVID-19 'vaccines' multifaceted harms by multiple pathophysiological mechanisms is, in 2025, beyond all reasonable doubt. However, the general public seem to be coming to this conclusion ahead of their highly qualified medical professional cohort. Despite this possibility, patients approaching doctors with questions about mandates, asking for the information aspect of informed consent, concerned about sudden morbidity in the face of a temporal relationship with the COVID jab are left exasperated. The accounts I am privy to, tell the story of seemingly unaware (or willfully blind) medics, dismissing claims and concerns, not looking at data or published journals that are deemed to go against the narrative, and going as far as gaslighting injured patients into diagnoses of 'functional' disorders and who are then offered a mental health referral and/or medications. And therein lies the first of many deviations from the guidance from GMP that the GMC make clear in their own wording of what is to be expected.

As far as jab uptake in the NHS is concerned, I have witnessed many staff unwilling to take the jab due to perceived ill-health, only to be happy to continue to 'follow orders' and administer them to patients despite personal refusal. I find this highly disturbing. It seems that an increasing proportion of NHS staff started to refuse the covid jabs after the second one as per ONS figures showing circa 70% refused the 2023/24 boosting booster.

Do we have an evolving increasing proportion of NHS staff as 'anti-vaxxers'? Or is there a new virus sweeping the workforce, namely burying head in the sand syndrome? Is it cognitive dissonance, willful blindness, mass formation, or all of the above?

Instead, it is the evil 'anti-vaxxer,' the antithesis of healthcare, that has faced the blame for the outcry of 'antivax' rhetoric, labelled 'tin hatters,' 'grifters,' 'quacks,' and 'conspiracy theorists.' Ethical, evidence-based safety concerns and evolving vaccine harm concerns are not only vilified, but the concerned doctors are actively targeted and disciplined, in short telling them to stay silent, or they are struck off by the GMC for raising such concerns based on their independent clinical practice experience. The irony particularly in the age of 'freedom to speak up guardianship' and 'whistleblower protection,' is that those, who at great personal, professional, financial, and reputational loss, voiced genuinely held concerns, faced the wrath and sickeningly eventual strike off from the medical register for such breaches. Labelled again ironically as villains that have 'negatively affected public confidence in the NHS' or more sickeningly labelled 'a risk,' 'a danger,' and accused of bringing the profession into disrepute. How can we move forward with one eye on the GMC guidance on GMP and Duty of Candour, when people who are FOLLOWING their guidance to the letter at great cost, and putting their patients first, are targeted by the regulators claim to 'protect the public' in their mission statement?

Either way, these things simply cannot be ignored regarding the Duty of Candour GMC guidelines. The public demand and expect an explanation of the paradox that is high vaccine promotion, low uptake. The key points from the

GMC duty of candour guideline are:

tell the person (or, where appropriate, their advocate, carer or family) when something has gone wrong;

apologise to the person (or, where appropriate, their advocate, carer or family);

offer an appropriate remedy or support to put matters right (if possible);

and finally explain fully to the person (or, where appropriate, their advocate, carer or family) the short and long term effects of what has happened.

These principles are inherent to our practice, even as to why we became a doctor in the first place, alongside the now defunct Hippocratic Oath. All four features above are essential for being open and honest that when 'something that goes wrong with their treatment or care causes, or has the potential to cause, harm or distress' the DUTY (not choice) applies. This is essential to maintaining integrity and more importantly trust in the profession that is ailing rapidly.

As a final note, I appeal to all those working in or around the field of medicine, with all the humility I have, despite my own blacklisting and the GMC tribunal as above: Doctors, Nurses, Paramedics, Allied healthcare staff, it is an unrelenting duty to follow the principles discussed in this book. When trust is lost, this can present a huge barrier to future compatibility. Covid has brought us into a new era of low confidence and enormous distrust in the medical profession. We need to openly and honestly admit error. We need to HELP, not gaslight, the poor victims and families of the rushed, unsafe, ineffective, novel gene injections that were pushed, coerced, mislabelled and misrepresented by each and every jab administrator and promotor. I implore you all to consider these points and do the right thing. The time to rebuild trust for the future of worldwide healthcare is long overdue. The alternative is to continue to let the insane run the asylum.

Many people who have been alert to COVID vaccine harm despite the repercussions continue to research the irrefutable data and studies are published week on week. There are ways to ameliorate the effects of the spike protein. Many are natural, cheap, safe and they work. It's time to admit where errors have been made, and to understand the pathophysiology of the spike protein and revert to why you joined the medical profession in the first place. Hopefully to help heal the body, soul, and mind of your patient through holistic, biopsychosocial, trustworthy, healthcare. Anything less is a failure of duty.

Chapter 23 - Final thoughts

In summary, my speaking out has led to suppression, censorship, character and reputational assassination, in addition to the end of my conventional medical career. I get asked if I had my time again, would I do the same?

There are things I would say differently – I allowed myself to be drawn into online spats and would certainly pause for reflection before responding online if indeed I would respond at all. I would know not to take the bait of anyone attempting to entrap me and I will never again use the phrase 'pile on.'

Beyond that, I regret nothing. When I became a doctor, I signed up to a professional oath to 'First do no harm' and a code of conduct underpinning good medical practice, ethics, and standards which I follow to the letter. I took an oath to become an ethical, safety focussed, evidence-based doctor, with attention to holistic patient centred care. They went after the children and pregnant women with unconscionable ease, and the collective lack of will to follow the DUTY of candour or speak out, at the time of writing in 2025, is sickening. I am proud of my stand. I have taken hit after hit, but I go back to my medical school interview response (Chapter One) as to why I wanted to become a doctor. I had no option to remain quiet. I had the same financial and career driven motivations to not speak out but, on balance, I wouldn't be able to sleep at night or look myself in the mirror if I had remained silent.

Most importantly, I want my children to know that their father was on the right side of history.

I find it ironic, that those that set out to damage me have simply amplified my presence and voice across all social media platforms. So, I would like to thank every troll for your abject failure in breaking me, silencing me, and ending my career. This is just the beginning. What doesn't kill you makes you stronger!

My prayers are for candour, integrity, and a united workforce among the medical community moving forwards, particularly in the NHS, once one of the crown jewels of the nation; I hope it can be again. God bless you all.

The End... of the beginning

Printed and bound by CPI Group (UK) Ltd, Croydon, CR0 4YY
03/09/2026
02208844-0001